Birkhäuser

Rinaldo B. Schinazi

From Classical to Modern Analysis

Rinaldo B. Schinazi
Department of Mathematics
University of Colorado
Colorado Springs, CO, USA

ISBN 978-3-030-06879-0 ISBN 978-3-319-94583-5 (eBook)
https://doi.org/10.1007/978-3-319-94583-5

Mathematics Subject Classification (2010): 26-01, 28-01, 40A05

Softcover re-print of the Hardcover 1st edition 2018

Printed on acid-free paper

This book is published under the imprint Birkhäuser, www.birkhauser-science.com by the registered company Springer Nature Switzerland AG
The registered company address is: Gewerbestrasse 11, 6330 Cham, Switzerland

To Blanquita

Preface

Why Another Analysis Book? There are many good undergraduate and graduate analysis books. There is, however, a rather large gap between undergraduate and graduate texts. The purpose of this book is to bridge this gap.

Typically, a graduate book covers measure theory and the Lebesgue integral. An undergraduate book does not. A book covering measure theory usually assumes that the reader is familiar with limits superior and inferior, arithmetic operations involving infinity, set theory, metric spaces, and some topology. Many students get into a measure theory course with only a vague notion of these concepts. Even more importantly, the level of abstraction the students have been exposed to is much lower than what is required to succeed in a graduate analysis course. As a result measure theory courses are much harder for the student than they need to be. The main goal of this book is to propose a gently rising path from undergraduate analysis (i.e., Classical Analysis) to measure theory (i.e., Modern Analysis).

Prerequisites We assume the reader of this text to have had an advanced calculus course at the level of Fitzpatrick (2006) or Schinazi (2011). In particular, the reader should be able to write and read short mathematical proofs. We also assume familiarity with the main results of the calculus of one variable. However, we do review most of what is needed as we go along and this book is largely self-contained.

What Courses Is This Book For? We have used these lecture notes for two semester long analysis courses for advanced undergraduate and beginning graduate students.

In the first semester we cover the first seven chapters. These chapters cover standard material such as number systems, convergence of sequences and series as well as more advanced topics such as limits superior and inferior, convergence of functions, and metric spaces.

In the second semester we cover Chapters 8 through 12 and selected topics from the last chapters. These last chapters are largely independent. The main purpose of the second course is an introduction to measure theory and the Lebesgue integral as well as a few applications.

Analysis is the theory of calculus. To be complete this theory requires measure theory and the Lebesgue integral. Most books at this level do not cover measure theory but choose to revisit calculus of several variables instead. In our experience, more calculus at this point does not do much for the student. Introducing measure theory on the other hand opens a whole new world.

Colorado Springs, CO, USA Rinaldo B. Schinazi

Contents

1 Number Systems 1
1 Real Numbers 1
1.1 Field Axioms 1
1.2 Order Axioms 4
Problems 6
1.3 The Fundamental Property of the Real Numbers 7
Problems 8
2 Natural Numbers 8
Problems 12
3 Rational Numbers 13
Problems 17

2 Sequences of Real Numbers 19
1 Convergence 19
2 Operations on Sequences 22
Problems 24
3 Monotone Sequences 26
4 Subsequences 27
5 Bolzano-Weierstrass Theorem 29
Problems 30
6 Cauchy Sequences 32
Problems 35
7 The Fundamental Property Revisited 36

3 Limits Superior and Inferior of a Sequence 39
1 Infinite Limits 39
2 Limit Superior and Limit Inferior 40
3 Operations on Limits 46
Problems 52

4 Numerical Series ... 55
1 First Properties ... 55
2 Positive Terms Series ... 57
Problems ... 61
3 Absolute Convergence ... 64
4 Conditional Convergence ... 68
5 Rearrangements ... 71
Problems ... 73

5 Convergence of Functions ... 77
1 Pointwise and Uniform Convergence ... 77
2 Uniform Convergence and Continuity ... 81
3 Uniform Convergence and Integration ... 83
4 Uniform Convergence and Differentiation ... 85
Problems ... 86
5 Polynomial Approximation of a Continuous Function ... 88
Problems ... 91
6 Continuity and Smoothness ... 92
6.1 M-Weierstrass Test ... 93
6.2 Nowhere Differentiability ... 94
Problems ... 97

6 Power Series ... 99
1 Radius of Convergence ... 99
2 Power Series and Uniform Convergence ... 101
3 Integrating and Differentiating Power Series ... 103
4 Convergence at Endpoints ... 107
5 Euler's Formula ... 110
Problems ... 113

7 Metric Spaces ... 115
1 Definition and Examples ... 115
1.1 Euclidean Spaces ... 115
1.2 Uniform Metric on Continuous Functions ... 118
1.3 Riemann Integral Metric on Continuous Functions ... 119
1.4 Euclidean Metric on a Space of Sequences ... 121
Problems ... 123
2 Sequences in Metric Spaces ... 124
2.1 Convergent Sequences ... 124
2.2 Cauchy Sequences ... 125
Problems ... 127
3 Complete Metric Spaces ... 127
3.1 Complete and Incomplete Euclidean Spaces ... 128
3.2 Complete and Incomplete Metrics on the Space of Continuous Functions ... 130
Problems ... 132

8 **Topology in a Metric Space** 137
1 Open and Closed Sets 137
Problems 144
2 Compactness 146
2.1 General Properties 146
2.2 Compact Sets in the Euclidean Spaces 147
2.3 A Closed Ball Which Is Not Compact 149
2.4 The Cantor Set 150
Problems 152

9 **Continuity on Metric Spaces** 155
1 Definition and Examples 155
2 Image and Inverse Image 157
3 Continuity and Open Sets 159
4 Continuity and Compactness 161
5 Continuity and Intervals 162
Problems 165

10 **Measurable Sets and Measurable Functions** 167
1 Measurable Sets 167
Problems 172
2 Measurable Functions 173
2.1 Definition and General Properties 173
2.2 Real Valued Functions 174
2.3 Operations on Measurable Functions 176
2.4 Sequences of Measurable Functions 177
2.5 Simple Functions 178
Problems 180

11 **Measures** 183
1 Definition and Examples 183
2 The Lebesgue Measure on the Real Line 189
2.1 Null Sets 190
2.2 Approximation of Lebesgue Measurable Sets and Functions 192
Problems 193

12 **The Lebesgue Integral** 197
1 The Integral of a Positive Function 197
1.1 The Integral of a Simple Function 197
Problems 203
1.2 The Integral of a Positive Real Function 204
1.3 The Monotone Convergence Theorem and Fatou's Lemma 206
1.4 Null Sets and Integration 211
Problems 214

2 The Integral of a Real Valued Function 216
2.1 The Definition of the Integral 216
2.2 Properties of the Integral 218
2.3 The Dominated Convergence Theorem 221
Problems 225

13 Integrals with Respect to Counting Measures 229
1 The Integral 229
2 Interchanging the Order of Summation 230
Problems 233

14 Riemann and Lebesgue Integrals 235
1 The Riemann Integral 235
2 Comparing the Riemann and Lebesgue Integrals 236
Problems 240

15 Modes of Convergence 243
1 The Lebesgue Integral Metric 243
1.1 The Lebesgue Integral Metric Is Complete 243
1.2 The Completion of the Riemann Integral Metric 247
Problems 248
2 Convergence in Measure 249
Problems 254
3 Uniform Integrability 257
Problems 265

References 267

Index 269

Chapter 1
Number Systems

1 Real Numbers

Mathematics is founded on axioms. These are properties that are given without proof, everything else can be proved using the axioms.

It is customary to first introduce natural numbers with the so-called Peano's axioms. Rational and real numbers are constructed using natural numbers, see for instance Krantz (1991). Following Apostol (1961), we take a different approach. Our axioms will define the real numbers. Natural and rational numbers will follow.

We will now list a set of axioms that define the real numbers. We will denote the set of real numbers by $\mathbb{R}$. The axioms define not only the real numbers but also the addition $x + y$ and multiplication xy of two real numbers x and y. We will also use the notation $x \cdot y$ for xy.

1.1 Field Axioms

Axiom 1. Commutativity. For any x and y in $\mathbb{R}$,

$$x + y = y + x \text{ and } xy = yx.$$

Axiom 2. Associativity. For any x, y, and z in $\mathbb{R}$,

$$(x + y) + z = x + (y + z) \text{ and } (xy)z = x(yz).$$

Axiom 3. Distributivity. For any x, y, and z in $\mathbb{R}$,

$$x(y + z) = xy + xz.$$

R. B. Schinazi, *From Classical to Modern Analysis*,
https://doi.org/10.1007/978-3-319-94583-5_1

Axiom 4. Identity elements. There exist two distinct real numbers 0 and 1 such that for every real number x

$$x + 0 = 0 + x = x \text{ and } x \cdot 1 = 1 \cdot x = x$$

Axiom 5. Inverses with respect to addition. For every real number x there exists a unique real number y such that $x + y = y + x = 0$. The number y is denoted by $-x$.

Axiom 6. Inverses with respect to multiplication. For every real number $x \neq 0$ there exists a unique real number y such that $xy = yx = 1$. The number y is denoted by $\frac{1}{x}$.

A set with the 6 properties above is called a **field**. These field axioms can be used to prove all the familiar arithmetic properties of the real numbers, see for instance Apostol (1961). We now give several examples. First we define the difference of two real numbers.

$$x - y = x + (-y).$$

Proposition 1.1

(i) *Let x be a real number, then $x \cdot 0 = 0 \cdot x = 0$.*
(ii) *Let x and y be real numbers. If $xy = 0$, then $x = 0$ or $y = 0$.*

Proof of Proposition 1.1
We first prove (i). By Axiom 4, $0 + 0 = 0$. Multiplying both sides by x we get,

$$x(0 + 0) = x \cdot 0.$$

By Axiom 3

$$x \cdot 0 + x \cdot 0 = x \cdot 0.$$

We know that $x \cdot 0$ like every real number has an inverse $-(x \cdot 0)$. We add $-(x \cdot 0)$ on both sides of the previous equality,

$$(x \cdot 0 + x \cdot 0) - x \cdot 0 = x \cdot 0 - x \cdot 0.$$

By Axiom 2 and Axiom 5,

$$x \cdot 0 = 0.$$

This completes the proof of (i).

We now turn to (ii). Assume that $xy = 0$. If $x = 0$, we are done. If $x \neq 0$, then x has an inverse and

$$\frac{1}{x}(xy) = \frac{1}{x} \cdot 0.$$

By Axiom 2 and (i) we get

$$(\frac{1}{x}x)y = 0.$$

By Axiom 6,

$$1 \cdot y = 0.$$

By Axiom 4, $y = 0$. The proof of (ii) is complete.

Proposition 1.2 *For all real numbers x and y,*

(i) $-(-x) = x$.
(ii) $-x = (-1)x$.
(iii) $(-x)y = -xy$.
(iv) $(-x)(-y) = xy$.

Proof of Proposition 1.2
We start with (i). Since $-(-x)$ is the inverse of $-x$ we have

$$-x + (-(-x)) = 0.$$

Since $-x$ is the inverse of x

$$-x + x = 0.$$

Hence, x and $-(-x)$ are both inverses of $-x$. Since inverses are unique, we have

$$-(-x) = x$$

and (i) is proved.

For (ii), by the distributive axiom

$$x + (-1)x = (1 + (-1))x.$$

By definition of -1, $(-1) + 1 = 0$. Thus,

$$x + (-1)x = 0 \cdot x = 0.$$

Therefore, $(-1)x$ is the unique inverse of x. That is $(-1)x = -x$. This proves (ii).

The proofs of (iii) and (iv) are left as exercises to the reader.

1.2 Order Axioms

The next axioms will be used to define the order relation in the real numbers.

We assume that the real numbers have a subset called the set of **strictly positive numbers**. We use the notation $x > 0$ to indicate that x is strictly positive.

Axiom 7 If $x > 0$ and $y > 0$, then $x + y > 0$ and $xy > 0$.

Axiom 8 For a real number x exactly one of the following three alternatives is true

$$x > 0, \qquad -x > 0, \qquad x = 0.$$

We define $x > y$ as $x - y > 0$. We define $x \geq y$ as $x > y$ or $x = y$. We define $x < y$ as $y > x$

We now list a few consequences of these axioms.

Proposition 1.3 *If* $x \neq 0$, *then* $x^2 > 0$.

Proof of Proposition 1.3
By Axiom 8 either $x > 0$ or $-x > 0$. If $x > 0$, then by Axiom 7, $x^2 = x \cdot x > 0$.

If $-x > 0$, then $(-x)(-x) > 0$. By Proposition 1.2, $(-x)(-x) = x \cdot x = x^2$. So for all $x \neq 0$, $x^2 > 0$. This completes the proof of Proposition 1.3.

- Since $1 = 1 \cdot 1 = 1^2$, Proposition 1.3 shows that $1 > 0$.

Proposition 1.4 *If* $x > 0$, *then* $\frac{1}{x} > 0$.

Proof of Proposition 1.4
By definition of the multiplicative inverse,

$$x\frac{1}{x} = 1.$$

Hence, $\frac{1}{x} \neq 0$ (why?). So either $\frac{1}{x} > 0$ or $-\frac{1}{x} > 0$. By contradiction assume $-\frac{1}{x} > 0$. Then,

$$x \cdot (-\frac{1}{x}) > 0.$$

By Proposition 1.2, the product above is -1. Hence, $-1 > 0$. But we cannot have at the same time $1 > 0$ and $-1 > 0$. We have a contradiction. Thus, $-\frac{1}{x} > 0$ is not true. Therefore, $\frac{1}{x} > 0$. Proposition 1.4 is proved.

Proposition 1.5

(i) *If* $x > y$ *and* $z > 0$*, then* $xz > yz$*.*
(ii) *If* $x > y$ *and* $z < 0$*, then* $xz < yz$*.*

Proof of Proposition 1.5
For (i), by definition $x > y$ means $x - y > 0$. By Axiom 7, $(x - y)z > 0$. Using the distributive axiom we have

$$xz - yz > 0$$

and therefore $xz > yz$. The proof of (i) is complete. The proof of (ii) is left to the reader.

Next we define the **absolute value** function. Let x be a real number, then

$$|x| = x \quad \text{if } x \geq 0$$

$$|x| = -x \text{ if } x < 0$$

Proposition 1.6 (The Triangle Inequality) *Let* a *and* b *be two real numbers. Then,*

$$|a + b| \leq |a| + |b|.$$

Proof of Proposition 1.6
If a and b are both positive or both negative, we just have $|a+b| = |a|+|b|$ (why?). The inequality holds.

Assume now that $a < 0 < b$. There are two cases.

If $a + b < 0$, then

$$|a + b| = -(a + b) = -a - b = |a| - |b| < |a| + |b|,$$

and the inequality holds.

If $a + b \geq 0$, then

$$|a + b| = (a + b) = -|a| + |b| < |a| + |b|,$$

and the inequality holds.

This completes the proof of Proposition 1.6.

Problems

1. Let $x \neq 0$ and $y \neq 0$.

(a) Show that

$$\frac{1}{-x} = -\frac{1}{x}.$$

(b) Show that

$$\frac{1}{xy} = \frac{1}{x}\frac{1}{y}.$$

2. For any real numbers x and y show that

(a) $(-x)y = -xy$.
(b) $(-x)(-y) = xy$.

3. Show that if $x < y$ and $y < z$, then $x < z$.

4. Show that if $x < y$ and $z < 0$, then $xz > yz$.

5.

(a) Show that if $0 \leq x < y$, then $x^2 < y^2$.
(b) Assume $x < y < 0$, what is the inequality between x^2 and y^2? Prove your claim.

6.

(a) Show that if $0 < x < y$, then

$$\frac{1}{x} > \frac{1}{y}.$$

(b) Assume $x < y < 0$, what is the inequality between $\frac{1}{x}$ and $\frac{1}{y}$? Prove your claim.

7. If a and b are both positive or both negative, show that $|a + b| = |a| + |b|$.

8. For two real numbers a and $b > 0$, show that $|a| < b$ if and only if $-b < a < b$.

9. Let a and b be two real numbers.

(a) Show that

$$|a| \leq |a - b| + |b|.$$

(b) Show that

$$|b| \leq |a - b| + |a|.$$

(c) Use (a) and (b) to show that

$$|a-b| \leq \big||a|-|b|\big|.$$

1.3 The Fundamental Property of the Real Numbers

The previous 8 axioms are algebraic properties. The last axiom we need to define the real numbers turns out not to be algebraic. We first need a few definitions.

A set A of real numbers is said to be **bounded above** if there exists a u such that if a is in A, then $a \leq u$. The number u is then called an **upper bound** for A. (Note that u need not be in A.)

A set A is said to have a **least upper bound** b if any upper bound u for A is such that $u \geq b$. If the set A has a least upper bound b, we denote it by

$$b = \sup A.$$

A set A of numbers is said to be **bounded below** if there exists an m such that if a is in A, then $a \geq m$. The number m is then called a **lower bound** for A. (Note that m need not be in A.)

A set A is said to have a **greatest lower bound** b if any lower bound m for A is such that $m \leq b$. If A has a greatest lower bound b we denote it by

$$b = \inf A.$$

We are now ready for our last axiom.

Axiom 9 (Fundamental Property of $\mathbb{R}$) A nonempty subset of $\mathbb{R}$ which is bounded above has a least upper bound.

- The Fundamental Property is equivalent to the following. A nonempty subset of $\mathbb{R}$ which is bounded below has a greatest lower bound.
- The reason why the Fundamental Property is fundamental is that it provides the *existence* of a real number (the least upper bound or the greatest lower bound of a given set) that may be missing in the rational numbers. For instance, we will use the fundamental property to show that $\sqrt{2}$ is a real number.

Example 1.1 Let $A = (-\infty, 1)$ (i.e., the set of real numbers strictly less than 1). Note that 1 is an upper bound of A. Hence, A is nonempty and bounded above. The Fundamental Property applies. The set A has a least upper bound. We now show it is 1.

Let $b < 1$ and

$$a = \frac{b+1}{2}.$$

Then, a is in A and $a > b$ (why?). Hence, b is not an upper bound of A (why?). Since 1 is an upper bound and anything less than 1 is not, 1 is the least upper bound of A.

Problems

1. Find the greatest lower bound of the set $[-2, 1)$.

2. Give an example of a set with no least upper bound.

3. Let $a > 0$ and $A = \{x \in \mathbb{R} : x > 0 \text{ and } x^2 < a\}$. Show that if $b^2 > a$, then b is an upper bound of A.

4. Show that a nonempty set which is bounded below has a greatest lower bound.

5.

(a) Let A be a subset of $\mathbb{R}$ and assume that A has a least upper bound $m = \sup A$. Show that for any $\epsilon > 0$ there exists an a in A such that

$$m - \epsilon < a \leq m.$$

(b) State and prove a property similar to (a) for $\inf A$.

6. Assume that A is a nonempty subset of real numbers. Assume that $A \subset B$ and that B has an upper bound. Show that $\sup A$ and $\sup B$ exist and that $\sup A \leq \sup B$.

7. State and prove a property analogous to the property in problem 6. for $\inf A$ and $\inf B$.

8. Let A and B be nonempty sets of real numbers with the following property. For all a in A and b in B we have $a \leq b$.

(a) Show that $\sup A$ and $\inf B$ exist and that $\sup A \leq \inf B$.
(b) Show that $\sup A = \inf B$ if and only if for every $\epsilon > 0$ there exist a in A and b in B such that $b - a < \epsilon$.

2 Natural Numbers

We start with a definition.

A set $A \subset \mathbb{R}$ is said to be **inductive** if

(i) 1 belongs to A.
(ii) if x belongs to A, then $x + 1$ belongs to A.

Note that there are many inductive sets in $\mathbb{R}$. In particular, the whole set $\mathbb{R}$ is inductive.

We define the **natural numbers** $\mathbb{N}$ as the intersection of all inductive sets. We have the following properties.

- The number 1 is the minimum of the set $\mathbb{N}$. That is, 1 belongs to $\mathbb{N}$ and if x belongs to $\mathbb{N}$, then $x \geq 1$.

We prove the claim. The number 1 belongs to every inductive set. Hence, it belongs to the intersection of all inductive sets which is $\mathbb{N}$.

Note that the set $A = \{x \in \mathbb{R} : x \geq 1\}$ is inductive (why?). Hence, $\mathbb{N}$ is a subset of A. Therefore, if $x \in \mathbb{N}$, then $x \in A$. That is, $x \geq 1$. This proves that 1 is the minimum of $\mathbb{N}$.

- $\mathbb{N}$ is an inductive set.

We now prove this claim. We already know that 1 belongs to $\mathbb{N}$. Assume now that x belongs to $\mathbb{N}$. Then, x belongs to every inductive set. Therefore, $x + 1$ belongs to every inductive set. Hence, $x + 1$ belongs to the intersection of all inductive sets. That is, $x + 1$ belongs to $\mathbb{N}$. Hence, $\mathbb{N}$ is an inductive set.

- **Principle of Mathematical Induction** If A is a subset of $\mathbb{N}$ and A is inductive, then A is all of $\mathbb{N}$.

The proof is easy. Since A is inductive and $\mathbb{N}$ is the intersection over all inductive sets, we must have $\mathbb{N} \subset A$. By assumption the reverse inclusion is also true. Hence, the two sets are equal.

Next we give a typical application of the Principle of Mathematical Induction.

Example 2.1 Show that for every natural n we have $2^n > n$.

Let A be defined by

$$A = \{n \in \mathbb{N} : 2^n > n\}.$$

Since $2^1 > 1$ we see that 1 belongs to A. Assume that n belongs to A. That is, $2^n > n$. Multiplying by 2 on both sides we get $2^{n+1} > 2n$. Since $n \geq 1$ we have $n + n \geq n + 1$ and therefore $2n \geq n + 1$. Hence,

$$2^{n+1} > 2n \geq n + 1.$$

Therefore, $2^{n+1} > n + 1$. That is, $n + 1$ belongs to A. This shows that A is an inductive set of natural numbers. By the Principle of Mathematical Induction A is all of $\mathbb{N}$. This proves the claim that for every natural n we have $2^n > n$.

Intuitively, the set $\mathbb{N}$ is $\{1, 1 + 1, (1 + 1) + 1, \dots\}$. We now give two properties along this line.

Proposition 2.1 Let $n > 1$ be a natural number. Then, $n - 1$ is also a natural number.

Proof of Proposition 2.1
Let

$$B = \{1\} \cup \{k \in \mathbb{N} : k - 1 \in \mathbb{N} \text{ and } k \in \mathbb{N}\}.$$

By definition 1 belongs to B. Assume that k belongs to B. Since k belongs to B, it belongs to $\mathbb{N}$. Therefore, $k + 1$ belongs to $\mathbb{N}$. That is, $k + 1$ and k belong to $\mathbb{N}$. Thus, $k + 1$ belongs to B. This proves that B is an inductive set. By the Principle of Mathematical Induction, B is all of $\mathbb{N}$. This shows that a natural number k is either 1 or is such that $k - 1$ is a natural number. This completes the proof of Proposition 2.1.

Proposition 2.2 *Let n be a natural number. Then, the interval $(n, n + 1)$ has no natural number.*

Proof of Proposition 2.2
Let S be the set of natural numbers n such that the interval $(n, n + 1)$ has no natural number. We will prove that S is an inductive set and is therefore all of $\mathbb{N}$.

Consider the set

$$A = \{1\} \cup \{n \in \mathbb{N} : n \geq 2\}.$$

Note that 1 is A. If x is in A so is $x + 1$ (why?). Hence, A is an inductive set. By the Principle of Mathematical Induction, A is all of $\mathbb{N}$. That is, a natural number is either 1 or is larger than or equal to 2. In particular, the interval $(1, 2)$ contains no natural number. This shows that 1 belongs to S.

Assume now that n belongs to S. By definition of S, the interval $(n, n+1)$ has no natural number. Assume by contradiction that $n + 1$ does not belong to S. That is, there exists a natural number x in the interval $(n + 1, n + 2)$. Since $x > n + 1 > 1$, by Proposition 2.1, $x - 1$ is a natural number. Observe that $x - 1$ is in $(n, n + 1)$. Hence, we have a natural number $x - 1$ in $(n, n + 1)$. This contradicts the induction hypothesis that n is in S. Hence, $n+1$ belongs to S. By the Principle of Mathematical Induction S is all of $\mathbb{N}$. The proof of Proposition 2.2 is complete.

The following is an important property of $\mathbb{N}$.

Theorem 2.1 (Well-Ordering Principle) Let A be a nonempty subset of $\mathbb{N}$. Then, A has a minimum. That is, there exists m **in** A such that if x is in A, then $x \geq m$.

Proof of Theorem 2.1
By contradiction assume that A has no minimum. Define the set B as

$$B = \{n \in \mathbb{N} : n < a \text{ for all } a \in A\}.$$

Since 1 is the minimum of $\mathbb{N}$, it cannot belong to A (otherwise 1 would be the minimum of A). Therefore, for all a in A, $a > 1$. That is, 1 belongs to B. Assume now that x belongs to B. By contradiction assume that $x + 1$ does not belong to B. Then, there exists a_1 in A such that $x + 1 \geq a_1$. Since A has no minimum there exists a_2 in A such that $a_2 < a_1$. Therefore, $x + 1 > a_2$. Since $x + 1$ and a_2 are natural numbers, we have $x \geq a_2$ (why?). But then x does not belong to B. We have a contradiction, $x+1$ must belong to B. By the Principle of Induction, B is all of $\mathbb{N}$. Since A is not empty we can find a in A. But B contains all natural numbers. In particular, $a \in B$ and therefore $a < a$. We have a contradiction. The set A does have a minimum. The proof of Theorem 2.1 is complete.

The Well-Ordering Principle is in fact equivalent to the Principle of Mathematical Induction, see the problems.

We now turn to an application of Theorem 2.1. First we introduce two sets of numbers.

The set of **integers** is defined by

$$\mathbb{Z} = \{\ldots, -2, -1, 0, 1, 2, 3 \ldots\}.$$

We denote the **rational numbers** by $\mathbb{Q}$. A rational number r can be represented by an ordered pair of integers (m, n) where $n \neq 0$. We write $r = \frac{m}{n}$. Next we use the Well-Ordering Principle to show that a rational number can be written as the ratio of two integers that are relatively prime.

Example 2.2 Show that every rational number $r > 0$ can be written as $r = \frac{a}{b}$ where a and b are natural numbers that are relatively prime. That is, the only divisor common to a and b is 1.

The rational number $r > 0$ is fixed. We define the set A by

$$A = \{n \in \mathbb{N} : rn \in \mathbb{N}\}.$$

Since $r > 0$ is a rational number there are natural numbers a and b such that $r = \frac{a}{b}$. Hence, $a = br$ and b belongs to A. This shows that A is not empty and therefore the Well-Ordering Principle applies. The set A has a minimum m. Since m is in A, mr is a natural number that we denote by k. Therefore, $r = \frac{k}{m}$. Once we show that k and m are relatively prime the proof will be complete. By contradiction assume that k and m have a common divisor $d > 1$. Then, there are natural numbers m' and k' such that

$$m = m'd \text{ and } k = k'd.$$

Since $rm = k$ we have $rm'd = k'd$ and therefore $rm' = k'$. In particular, m' belongs to A. Since $d > 1$ we have $m'd > m'$. That is, $m > m'$. But m is the minimum of A! Hence, m' cannot belong to A. We have a contradiction. Hence, k and m must be relatively prime. This concludes the proof.

The following simple result will turn out to be quite useful.

Theorem 2.2 (Archimedean Property) *For any real number a there is a natural number n such that $n > a$.*

Proof of Theorem 2.2
We do a proof by contradiction. Assume that the Archimedean property is not true. Then, there exists a real number a such that for all natural numbers n we have $n \leq a$. Thus, $\mathbb{N}$ which is a nonempty subset of real numbers is bounded above by a. The Fundamental Property applies. The set $\mathbb{N}$ has a least upper bound m. Hence, $m-1$ is not an upper bound of $\mathbb{N}$. Thus, there exists a natural number n such that $n > m-1$. Therefore,

$$n + 1 > m.$$

That is, $n + 1$ which is a natural number is larger than m which is an upper bound of the natural numbers. We have our contradiction. The Archimedean property must hold. This completes the proof of Theorem 2.2

Example 2.3 Show that for any real number $x > 0$ there exists a natural number m such that $x \leq m < x + 1$.

If $0 < x \leq 1$, then $x \leq 1 < x + 1$. The property holds for $m = 1$.
Assume now that $x > 1$. Let

$$A = \{n \in \mathbb{N} : n \geq x\}.$$

By the Archimedean Property A is not empty. By the Well-Ordering Principle A has a minimum m. Therefore, $m - 1$ is not in A. Note that for an element to be in A it has to meet two conditions. It has to be in $\mathbb{N}$ and it has to be larger than x. Hence, in order for $m - 1$ not to be in A we must have that $m - 1$ is not a natural number or that $m - 1 < x$. Since $m \geq x > 1$, $m - 1$ is a natural number. Hence, we must have $m - 1 < x$. Thus,

$$x \leq m < x + 1.$$

Problems

1. Show that for every natural number n,

$$1 + 2 + \cdots + n = \frac{n(n+1)}{2}.$$

2. Show that for every natural number n,

$$1^2 + 2^2 + \cdots + n^2 = \frac{n(n+1)(2n+1)}{6}.$$

3. Let $x > 0$ and $\epsilon > 0$ be two real numbers. Show that there exists an integer $n \geq 0$ such that

$$n\epsilon \leq x < (n+1)\epsilon.$$

(Show that the set $A = \{n \in \mathbb{N} : n\epsilon > x\}$ has a minimum.)

4. Let n be a natural number. Let A be the set of natural numbers larger or equal to 2 that divide n.

(a) Show that A has a minimum p.
(b) Show that the number p is prime (i.e., the only divisors of p are 1 and p).
(c) Show that every natural number has at least one prime divisor.

5. Let A be the collection of all inverses of natural numbers. Find the greatest lower bound of A.

6. Let $a \geq 0$. Assume that $a < 1/n$ for every natural n. Show that $a = 0$.

7. In this problem we prove that the Well-Ordering Principle implies the Principle of Mathematical Induction. Putting this together with Theorem 2.1 proves that the two principles are equivalent.

Assume the Well-Ordering Principle holds. Let A be an inductive set of natural numbers. Let B be the complement of A (i.e., all the natural numbers not in A). We are going to prove that B is empty. That will prove that A is all of $\mathbb{N}$.

(a) Assume that $B \neq \emptyset$. Show that B has a minimum m.
(b) Show that $m \geq 2$.
(c) Show that $m - 1$ is in A.
(d) Show that m is in A. Find the contradiction and conclude.

3 Rational Numbers

As defined above a rational number can be represented as a ratio of two integers. It is easy to check that the first 8 axioms listed above hold for the set of rational numbers $\mathbb{Q}$. That is, $\mathbb{Q}$ is an ordered field. Axiom 9 (i.e., the Fundamental Property) however does not hold for $\mathbb{Q}$, see the problems.

Proposition 3.1 *The equation $x^2 = 2$ has no solution in the rational numbers.*

We will show below that the equation $x^2 = 2$ has solutions in $\mathbb{R}$. This together with Proposition 3.1 shows that $\mathbb{Q}$ is strictly included in $\mathbb{R}$.

Proof of Proposition 3.1
We will need the following elementary facts. A natural number m which is divisible by 2 (i.e., $m = 2p$ for some natural p) is called an even number. A natural number which is not divisible by 2 (i.e., $m = 2p + 1$ for some natural p) is called an odd

number. A natural number cannot be both even and odd. In particular, m is even if and only if m^2 is even. See for instance Section 1.2 in Schinazi (2011).

We do a proof by contradiction. Assume that there exists a rational number r such that $r^2 = 2$. We can assume $r > 0$ because if r is a solution so is $-r$. Therefore, there are natural numbers a and b such that $r = \frac{a}{b}$. By Example 2.2 we can pick a and b to be relatively prime. Since $r^2 = 2$ we have $a^2 = 2b^2$. That is, 2 divides a^2. It is not difficult to see that 2 must then also divide a. Hence, there is a natural number a' such that $a = 2a'$. Therefore,

$$(2a')^2 = 2b^2$$

and 2 divides b^2. Hence, 2 divides b. Thus, 2 is a common divisor of a and b. They are not relatively prime. We have a contradiction. There is no rational number which is a solution to $x^2 = 2$. This completes the proof of Proposition 3.1.

Next we show that the equation $x^2 = 2$ does have a solution in $\mathbb{R}$. The reader should note in the proof below the critical importance of the Fundamental Property of $\mathbb{R}$.

Proposition 3.2 *Let $a \geq 0$ be a real number. The equation $x^2 = a$ has exactly one positive solution in $\mathbb{R}$. This positive solution is denoted by $\sqrt{a}$.*

Proof of Proposition 3.2
Note that $x^2 = 0$ has the unique solution $x = 0$ (why?). Hence, $\sqrt{0} = 0$. Similarly, $\sqrt{1} = 1$. For the rest of the proof we assume $a \neq 0$ and $a \neq 1$.

If $x^2 = a$ and $y^2 = a$, then

$$x^2 - y^2 = (x - y)(x + y) = 0.$$

Either $x = y$ or $x = -y$. This shows that there is at most one positive solution to the equation $x^2 = a$. That is, if there is a positive solution it must be unique.

We now turn to the existence of the solution. Fix $a > 0$ and let

$$A = \{x \in \mathbb{R} : x > 0 \text{ and } x^2 < a\}.$$

Note that if $0 < a < 1$, then $a^2 < a$ and a is in A. If $a > 1$, then $1^2 = 1 < a$ and 1 is in A. Hence, A is not empty. Note that

$$(a + 1)^2 = a^2 + 2a + 1 > a.$$

Hence, if x is in A, then $x^2 < a < (a + 1)^2$. Thus, $x < a + 1$ (why?) and A is bounded above by $a+1$. According to the Fundamental Property of $\mathbb{R}$, A has a least upper bound m. Since the numbers in A are strictly positive we must have $m > 0$.

The number m will provide the positive solution of the equation $x^2 = a$ we are looking for. In order to prove this we are going to show that we cannot have $m^2 < a$ and we cannot have $m^2 > a$. The only possibility left will be $m^2 = a$.

Assume first that $m^2 < a$ and we will show that this leads to a contradiction. For two numbers a and b we use the notation $\min(a, b)$ to denote the smallest of a and b. Let

$$\epsilon = \min(1, \frac{a - m^2}{2(2m + 1)}).$$

Thus, ϵ is less than 1 but strictly larger than 0. We now compute

$$\begin{aligned}(m + \epsilon)^2 &= m^2 + 2m\epsilon + \epsilon^2 \\ &\leq m^2 + 2m\epsilon + \epsilon \\ &= m^2 + (2m + 1)\epsilon\end{aligned}$$

where the inequality $\epsilon^2 \leq \epsilon$ comes from the fact that $\epsilon \leq 1$. Now we use that

$$\epsilon \leq \frac{a - m^2}{2(2m + 1)}$$

to get

$$\begin{aligned}(m + \epsilon)^2 &\leq m^2 + (2m + 1)\frac{a - m^2}{2(2m + 1)} \\ &= m^2 + \frac{a - m^2}{2} \\ &= \frac{a + m^2}{2}.\end{aligned}$$

Since $m^2 < a$, $\frac{a+m^2}{2} < a$. Therefore, $m + \epsilon$ belongs to A. The number $m + \epsilon$ is strictly larger than m which is an upper bound of A. Hence, $m + \epsilon$ cannot belong to A. We have a contradiction. The assumption $m^2 < a$ cannot hold.

Next we show that the assumption $m^2 > a$ also leads to a contradiction. Set

$$\epsilon = \frac{m^2 - a}{4m}.$$

Then,

$$\begin{aligned}(m - \epsilon)^2 &= m^2 - 2m\epsilon + \epsilon^2 \\ &> m^2 - 2m\epsilon \\ &= m^2 - 2m\frac{m^2 - a}{4m}\end{aligned}$$

$$= \frac{m^2 + a}{2}$$

$$> a.$$

That is, by picking $\epsilon > 0$ small enough we get that $(m - \epsilon)^2 > a$. This shows that $m - \epsilon$ is an upper bound of A (why?). But $m - \epsilon$ is strictly smaller than m which is the least upper bound of A. Thus, $m - \epsilon$ cannot be an upper bound of A. We have a contradiction. Therefore, we cannot have $m^2 > a$.

Since we cannot have $m^2 < a$ nor $m^2 > a$ we must have $m^2 = a$. That is, m is the unique positive solution of the equation $x^2 = a$. This completes the proof of Proposition 3.2.

Even though $\mathbb{Q}$ is strictly included in $\mathbb{R}$, the result below shows that rational numbers are everywhere!

Theorem 3.1 (Density of the Rationals) *Let $a < b$ be two real numbers. Then, there exists a rational number r such that $a < r < b$.*

In words, there exists always a rational number between two real numbers. The rational numbers are said to be **dense** in $\mathbb{R}$.

Proof of Theorem 3.1
We will prove the theorem in the particular case $0 < a < b$. The other cases are an easy consequence of this one.

By the Archimedean property there exists a natural p such that $p > \frac{1}{b-a}$. In Example 2.1 we proved that $2^n > n$ for every natural number n. Hence,

$$2^p > \frac{1}{b-a}.$$

Therefore, $b2^p - a2^p > 1$. By Example 2.3, there exists a natural number m such that

$$a2^p \leq m < a2^p + 1.$$

Since $a2^p + 1 < b2^p$,

$$a \leq \frac{m}{2^p} < b.$$

Let $r = \frac{m}{2^p}$, then r is a rational number and is in $[a, b)$. If $r > a$ we are done. We have found a rational number strictly between a and b. If $r = a$, then a is a rational number. Hence, $r' = a + \frac{1}{2^p}$ is also a rational number and is strictly between a and b. This completes the proof of Theorem 3.1.

An **irrational** number is a real number which is not rational.

Theorem 3.2 (Density of the Irrationals) *Let $a < b$ be two real numbers. Then, there exists an irrational number s such that $a < s < b$.*

The proof is a consequence of Theorem 3.1 and is left to the reader, see the problems.

Problems

1. In this problem we show that the irrational numbers are dense in $\mathbb{R}$. Let $a < b$ be two real numbers.

(a) Show that there is a rational number r strictly between a and b.
(b) Show that there is a natural number n such that $\sqrt{2} < n(b-r)$.
(c) Let

$$s = r + \frac{\sqrt{2}}{n}.$$

Show that s is irrational.
(d) Show that s strictly between a and b.

2. In Proposition 3.2 we proved that $x^2 = a$ has exactly one positive solution in $\mathbb{R}$.. This positive solution is denoted by $\sqrt{a}$.

(a) For any real x show that $\sqrt{x^2} = |x|$.
(b) Show that if $0 \leq x < y$, then $\sqrt{x} < \sqrt{y}$.
(c) Show that if $x \geq 0$ and $y \geq 0$, then

$$\sqrt{xy} \leq \frac{x+y}{2}.$$

(Start with $(\sqrt{x} - \sqrt{y})^2 \geq 0$.)

3. Let B be the following set of rational numbers.

$$B = \{r \in \mathbb{Q} : r < \sqrt{2}\}.$$

(a) Show that $m = \sup B$ exists in $\mathbb{R}$.
(b) Show that $m = \sqrt{2}$.
(c) Conclude that the Fundamental Property does not hold for $\mathbb{Q}$.

Chapter 2
Sequences of Real Numbers

1 Convergence

A sequence of real numbers is a function defined on a subset of positive integers with values in the real numbers. Instead of denoting the sequence by a (as we would for a function) we will use the notation (a_n). Note that a_p denotes the value of the sequence at p and (a_n) denotes the whole sequence.

The sequence (a_n) is said to **converge** to a real number ℓ if for every $\epsilon > 0$, there is a natural number N (that usually depends on ϵ) such that if $n \geq N$, then

$$|a_n - \ell| < \epsilon.$$

The number ℓ is said to be the limit of the sequence (a_n). Since the limit for a sequence will always be taken at positive infinity we will abbreviate the notation $\lim_{n\to\infty} a_n = \ell$ and write instead

$$\lim_n a_n = \ell.$$

Example 1.1 Consider a constant sequence (a_n) defined by $a_n = c$, for all $n \geq 1$, where c is a constant. Take any $\epsilon > 0$, take $N = 1$, then for every $n \geq N$ we have $|a_n - c| = 0 < \epsilon$. Therefore, (a_n) converges to c.

Example 1.2 For $n \geq 1$, let $a_n = 1/n$. Let $\epsilon > 0$, by the Archimedean property there is a natural $N > 1/\epsilon$. For $n \geq N$, we have $1/n \leq 1/N < \epsilon$ (why?). Thus,

$$\frac{1}{n} = |a_n - 0| < \epsilon \text{ for all } n \geq N.$$

This proves that (a_n) converges to 0.

R. B. Schinazi, *From Classical to Modern Analysis*,
https://doi.org/10.1007/978-3-319-94583-5_2

Example 1.3 Assume that the sequence (a_n) converges to some ℓ. Let k be a constant. Then the sequence (ka_n) converges to $k\ell$.

Note that if $k = 0$, then (ka_n) is identically 0 and therefore converges to 0. The property holds for $k = 0$. Assume now that $k \neq 0$. Let $\epsilon > 0$. There exists N such that for $n \geq N$ we have

$$|a_n - \ell| < \frac{\epsilon}{|k|}.$$

Thus,

$$|ka_n - k\ell| < \epsilon,$$

for $n \geq N$. Hence, (ka_n) converges to $k\ell$.

Next we show that the limit of a sequence is unique.

Proposition 1.1 *Assume that the sequence (a_n) converges to a and converges to b. Then, we must have $a = b$.*

Proof of Proposition 1.1
Let $\epsilon > 0$. There exists N_1 such that if $n \geq N_1$, then $|a_n - a| < \frac{\epsilon}{2}$. Similarly, there exists N_2 such that if $n \geq N_2$, then $|a_n - b| < \frac{\epsilon}{2}$. Let N be the largest of N_1 and N_2. Then, for $n \geq N$ we have

$$|a - b| = |a - a_n + a_n - b| \leq |a - a_n| + |a_n - b| < \frac{\epsilon}{2} + \frac{\epsilon}{2} = \epsilon.$$

That is, $|a - b| < \epsilon$ for every $\epsilon > 0$. This is only possible if $a = b$ (why?). This completes the proof of Proposition 1.1.

A sequence (a_n) is said to be **bounded** if there exists a real k such that $|a_n| < k$ for every $n \geq 1$.

Proposition 1.2 *Assume that the sequence (a_n) is convergent. Then, (a_n) is bounded.*

Proof of Proposition 1.2
Assume that the sequence (a_n) converges to some limit a. Take $\epsilon = 1$, since (a_n) converges to a there is a natural N such that if $n \geq N$, then $|a_n - a| < \epsilon = 1$. Thus, by the triangle inequality we have

$$|a_n| = |a_n - a + a| \leq |a_n - a| + |a| < 1 + |a| \text{ for all } n \geq N.$$

The inequality above shows that the sequence is bounded for $n \geq N$. We now take care of $n < N$. A finite set of real numbers always has a maximum and a minimum. Thus, let

$$m = \max\{|a_1|, |a_2|, \ldots, |a_{N-1}|\}$$

and let $k = \max(m, 1 + |a|)$. Then, we claim that

$$|a_n| \leq k \text{ for all } n \geq 1.$$

For if $n < N$, then $|a_n| \leq m \leq k$, while if $n \geq N$, then $|a_n| < |a| + 1 \leq k$. This completes the proof of Proposition 1.2.

As the next example shows a bounded sequence need not converge!

Example 1.4 For $n \geq 1$, let $a_n = (-1)^n$. Then $|a_n| = 1$ for every $n \geq 1$. Hence, (a_n) is bounded. However, (a_n) does not converge. It keeps jumping between -1 and 1. We will give a formal proof below.

Example 1.5 For $n \geq 1$, let $a_n = n$. Then, (a_n) is not bounded (why?). Hence, by Proposition 1.2, (a_n) cannot converge.

The following so-called *Squeezing Principle* may be useful in showing convergence and finding the limit of a sequence.

Proposition 1.3 *Assume that the three sequences (a_n), (b_n), and (c_n) are such that for every $n \geq 1$*

$$a_n \leq b_n \leq c_n.$$

Assume also that the sequences (a_n) and (c_n) converge to the same limit ℓ. Then (b_n) converges to ℓ as well.

Proof of Proposition 1.3
We will use repeatedly the elementary fact that for $a \geq 0$, $|x| < a$ if and only if $-a < x < a$.

Since (a_n) and (c_n) converge to ℓ, for any $\epsilon > 0$ there are naturals N_1 and N_2 such that if $n \geq N_1$, then $|a_n - \ell| < \epsilon$ and if $n \geq N_2$, then $|c_n - \ell| < \epsilon$. Let $N = \max(N_1, N_2)$. For $n \geq N$, we have

$$\ell - \epsilon < a_n < \ell + \epsilon \text{ and } \ell - \epsilon < c_n < \ell + \epsilon.$$

Therefore,

$$\ell - \epsilon < a_n \leq b_n \leq c_n < \ell + \epsilon.$$

Hence, for all $n \geq N$

$$\ell - \epsilon < b_n < \ell + \epsilon.$$

That is, (b_n) converges to ℓ. This completes the proof of Proposition 1.3.

Example 1.6 Assume that (a_n) converges to 0 and that (b_n) is a bounded sequence. Then, the sequence $(a_n b_n)$ converges to 0.

Since (b_n) is bounded there is k such that $|b_n| < k$ for all n. Then,

$$0 \leq |a_n b_n| \leq k|a_n|.$$

That is, the sequence $|a_n b_n|$ is squeezed between the constant sequence 0 and the sequence $k|a_n|$. But they both converge to 0 (why?). By the Squeezing Principle the sequence $(|a_n b_n|)$ converges to 0 and so does $(a_n b_n)$ (why?).

2 Operations on Sequences

We now state some elementary properties of convergent sequences.

Proposition 2.1 *Assume that the sequences (a_n) and (b_n) converge to a and b, respectively. Then,*

(i) For any constant k, the sequence (ka_n) converges to ka.
(ii) The sequence $(a_n + b_n)$ converges to $a + b$.
(iii) The sequence $(a_n b_n)$ converges to ab.
(iv) If $b_n \neq 0$ for all $n \geq 1$ and $b \neq 0$, then the sequence $(\frac{a_n}{b_n})$ converges to $\frac{a}{b}$.

Proof of Proposition 2.1
Note that (i) was proved in Example 1.3.

We prove (ii). There are natural numbers N_1 and N_2 such that if $n \geq N_1$ we have $|a_n - a| < \frac{\epsilon}{2}$ and if $n \geq N_2$ we have $|b_n - b| < \frac{\epsilon}{2}$. Take $N = \max(N_1, N_2)$ so that both statements above hold for $n \geq N$ and use the triangle inequality to get

$$\begin{aligned}|(a_n + b_n) - (a + b)| =&|a_n - a + b_n - b|\\ \leq&|a_n - a| + |b_n - b|\\ <&\frac{\epsilon}{2} + \frac{\epsilon}{2} = \epsilon.\end{aligned}$$

This proves (ii).

We now deal with (iii). Since (a_n) converges it is bounded. If $b = 0$, then by Example 1.6, $(a_n b_n)$ converges to 0. This proves (iii) when $b = 0$.

Assume now that $b \neq 0$. Take $\epsilon > 0$, there is N_1 such that if $n \geq N_1$, then

$$|a_n - a| < \frac{\epsilon}{2|b|}.$$

There exists $k > 0$ such that $|a_n| < k$ for all $n \geq 1$ (why?). There is also N_2 such that if $n \geq N_2$, then

$$|b_n - b| < \frac{\epsilon}{2k}.$$

Hence, if $n \geq \max(N_1, N_2)$

$$\begin{aligned}|a_n b_n - ab| =&|a_n(b_n - b) + a_n b - ab|\\ \leq&|a_n||b_n - b| + |b||a_n - a|\\ <&k\frac{\epsilon}{2k} + |b|\frac{\epsilon}{2|b|} = \epsilon.\end{aligned}$$

This completes the proof of (iii).

To prove (iv) it is enough to prove that $(1/b_n)$ converges to $1/b$ and then use (iii) (why?). We start with

$$|\frac{1}{b_n} - \frac{1}{b}| = |\frac{b_n - b}{b_n b}|.$$

The main difficulty is to show that the sequence (b_n) is bounded away from 0. We show this now. Since (b_n) converges to $b \neq 0$, there is a natural number N_1 such that if $n \geq N_1$, then

$$|b_n - b| \leq \frac{|b|}{2}.$$

By the triangle inequality we get, for $n \geq N_1$,

$$\begin{aligned}|b| =&|b - b_n + b_n|\\ \leq&|b - b_n| + |b_n|\\ <&\frac{|b|}{2} + |b_n|.\end{aligned}$$

Hence, $|b_n| > |b|/2$ for all $n \geq N_1$. That is, (b_n) is bounded away from 0 for n large enough. Thus,

$$\frac{1}{|b_n b|} < \frac{2}{b^2} \text{ for all } n \geq N_1.$$

On the other hand, since (b_n) converges to b, for any $\epsilon > 0$ there is N_2 such that if $n \geq N_2$, then

$$|b_n - b| < \frac{\epsilon b^2}{2}.$$

Therefore, for $n \geq \max(N_1, N_2)$ we have

$$|\frac{1}{b_n} - \frac{1}{b}| =|\frac{b_n - b}{b_n b}|$$

$$= \frac{|b_n - b|}{|b_n b|} |$$

$$< \frac{\epsilon b^2}{2} \frac{2}{b^2} = \epsilon.$$

That is, $(\frac{1}{b_n})$ converges to $\frac{1}{b}$. This completes the proof of Proposition 2.1.

Proposition 2.2 *Assume that the sequences (a_n) and (b_n) converge. Assume also that for all $n \geq 1$ we have $a_n \leq b_n$. Then,*

$$\lim_n a_n \leq \lim_n b_n.$$

- If $a_n < b_n$ for all $n \geq 1$, then Proposition 2.2 still applies (why?) and we get $\lim_n a_n \leq \lim_n b_n$. However, the strict inequality $\lim_n a_n < \lim_n b_n$ need not be true. See the problems.

Proof of Proposition 2.2
We do a proof by contradiction. Assume that (a_n) and (b_n) converge to a and b with $a > b$. Let $\epsilon = \frac{a-b}{2} > 0$. There exists N_1 such that if $n \geq N_1$, then

$$|a_n - a| < \epsilon.$$

In particular, $a_n > a - \epsilon = \frac{a+b}{2}$, if $n \geq N_1$. On the other hand, there exists N_2 such that if $n \geq N_2$, then

$$|b_n - b| < \epsilon.$$

In particular, $b_n < b + \epsilon = \frac{a+b}{2}$, if $n \geq N_2$. Thus, if $n > \max(N_1, N_2)$, then

$$b_n < \frac{a+b}{2} < a_n$$

contradicting the assumption $a_n \leq b_n$ for all n. This completes the proof of Proposition 2.2.

Problems

1. Let x be a real number such that $0 \leq x \leq \epsilon$ for every $\epsilon > 0$. Show that $x = 0$.

2.

(a) Show that if a_n converges to a, then $|a_n|$ converges to $|a|$.
(b) Is it true that if $|a_n|$ converges, then a_n converges? Prove it or give a counter-example.

3. Show that a sequence (a_n) converges to 0 if and only if the sequence $(|a_n|)$ converges to 0.

4. Let (a_n) be a sequence. For $n \geq 1$, let $b_n = |a_n - \ell|$ where ℓ is a real number. Show that (a_n) converges to ℓ if and only if (b_n) converges to 0.

5. Assume that $|a_n - a| \leq 1/n$ for all $n \geq 1$. Show that (a_n) converges to a.

6. For $n \geq 1$, let $a_n = \frac{(-1)^n}{n}$. Does (a_n) converge?

7. Assume that (a_n) converges to ℓ.

(a) For every $n \geq 0$, let $b_n = a_{n+1}$. Show that (b_n) converges.
(b) For every $n \geq 2$, let $c_n = a_{n-1}$. Show that (c_n) converges.
(c) State and prove a claim that generalizes (a) and (b).

8. Assume that the sequences (a_n) and $(a_n + b_n)$ converge. Show that (b_n) converges.

9. Let A be a subset in $\mathbb{R}$ that has a greatest lower bound k.

(a) For every $n \geq 1$ show that there exists a_n in A such that

$$a_n < k + \frac{1}{n}.$$

(b) Show that (a_n) converges to k.
(c) Assume that A has a least upper bound m. Show that there exists a sequence (b_n) in A such that (b_n) converges to m.

10. Assume that the sequence (a_n) is positive and converges to a limit ℓ. Show that $(\sqrt{a_n})$ converges to $\sqrt{\ell}$.

11. Let $a_n = n^{1/n}$ and $b_n = a_n - 1$ for $n \geq 1$.

(a) Show that $b_n > 0$ for all $n \geq 2$.
(b) Show that

$$n = (b_n + 1)^n \geq \frac{n(n-1)}{2} b_n^2.$$

(c) Use (a) and (b) to get

$$0 < b_n \leq \sqrt{\frac{2}{n-1}}.$$

(d) Show that $n^{1/n}$ converges to 1.

12. Assume that $a_n < b_n$ for all naturals n. Assume that (a_n) and (b_n) converge to a and b, respectively. Is it true that $a < b$? Prove it or give a counter-example.

13. Let $a_n \neq 0$ for every $n \geq 1$. Assume also that (a_n) converges to $a \neq 0$. Show that there is a $b > 0$ such that $|a_n| > b$ for every $n \geq 1$.

14. Assume that (a_n) converges to ℓ. Let $b < \ell$. Show that there is an N such that if $n \geq N$, then $b < a_n$.

15. Let (a_n) be a sequence. Define the sequence (b_n) by

$$b_n = \frac{1}{n}(a_1 + a_2 + \cdots + a_n),$$

for $n \geq 1$.

(a) Show that if (a_n) converges to 0 so does (b_n).
(b) Use (a) to show that if (a_n) converges, then (b_n) converges to the same limit.
(c) Give an example for which (b_n) converges but (a_n) does not.

3 Monotone Sequences

First some definitions.

A sequence (a_n) is said to be **increasing** if for every natural n we have $a_{n+1} \geq a_n$.

A sequence a_n is said to be **decreasing** if for every natural n we have $a_{n+1} \leq a_n$.

A sequence which is increasing or decreasing is said to be **monotone**.

Monotone sequences have nice properties. However, most sequences are not monotone.

A sequence (a_n) is said to be **bounded below** if there is a real number m such that $a_n > m$ for all $n \geq 1$.

A sequence (a_n) is said to be **bounded above** if there is a real number k such that $a_n < k$ for all $n \geq 1$.

We are now ready to state an important result for monotone sequences.

Theorem 3.1 *An increasing sequence converges if and only if it is bounded above. A decreasing sequence converges if and only if it is bounded below.*

Proof of Theorem 3.1
We prove the statement about increasing sequences, the other one is similar and is left as an exercise.

Assume that the sequence (a_n) converges, then we know it must be bounded and therefore bounded above. This proves one implication.

For the other implication assume that the increasing sequence (a_n) is bounded above by a number m. Let A be the set

$$A = \{a_n, n \in \mathbb{N}\}.$$

The set A has an upper bound m and is not empty. The Fundamental Property of $\mathbb{R}$ applies. The set A has a least upper bound ℓ. We are now going to show that the sequence (a_n) converges to ℓ. Take any $\epsilon > 0$, since ℓ is the *least* upper bound, then $\ell - \epsilon$ cannot be an upper bound. That is, there is an element in A strictly larger than $\ell - \epsilon$. Therefore, there is N such that $a_N > \ell - \epsilon$. Since (a_n) is increasing we have that if $n \geq N$, then

$$a_n \geq a_N > \ell - \epsilon.$$

Hence,

$$\ell - a_n < \epsilon.$$

For all $n \geq 1$, $\ell \geq a_n$, therefore

$$|a_n - \ell| = \ell - a_n < \epsilon \text{ for all } n \geq N.$$

That is, (a_n) converges to ℓ. This completes the proof of Theorem 3.1.

Example 3.1 Let c be in $(0, 1)$. For $n \geq 1$, let $a_n = c^n$. Show that (a_n) converges to 0.

Since $0 < c < 1$ we have $c^n c < c^n$ for every $n \geq 1$. That is, $a_{n+1} < a_n$. The sequence (a_n) is strictly decreasing. It is also a positive sequence. Therefore, it is bounded below by 0. By Theorem 3.1, the sequence (a_n) converges to some ℓ.

We now show that $\ell = 0$. For $n \geq 1$, let $b_n = a_{n+1}$. Then, (b_n) converges to the same limit ℓ (why?). However, $b_n = ca_n$ and since (a_n) converges to ℓ we know that (ca_n) converges to $c\ell$. Hence, (b_n) converges to ℓ and to $c\ell$. Since the limit is unique we have $\ell = c\ell$. Therefore, either $c = 1$ or $\ell = 0$. Since $c < 1$ we must have $\ell = 0$.

4 Subsequences

Let $1 \leq j_1 < j_2 < \cdots < j_n < \ldots$ be a strictly increasing sequence of natural numbers. Let (a_n) be a sequence of real numbers, then

$$(a_{j_n}) = \{a_{j_1}, a_{j_2}, \ldots, a_{j_n}, \ldots\}$$

defines a new sequence of real numbers. The sequence (a_{j_n}) is said to be a **subsequence** of the sequence (a_n).

- Here are a few examples of strictly increasing sequences (j_n) of natural numbers. For all $n \geq 1$, let $j_n = n$, $j_n = 2n$, $j_n = 2n + 1$, or $j_n = 2^n$.

- If (j_n) is a strictly increasing sequence of natural numbers, then we have $j_n \geq n$ for all $n \geq 1$ (why?).

Theorem 4.1 *A sequence (a_n) converges to a if and only if all the subsequences of (a_n) converge to a.*

Proof of Theorem 4.1
Assume that (a_n) converges to a and let (a_{j_n}) be a subsequence of (a_n). For any $\epsilon > 0$, there is N such that if $n \geq N$, then $|a_n - a| < \epsilon$. Since $j_n \geq n$ for all $n \geq 1$, if $n \geq N$, then $j_n \geq n \geq N$. That is, there exists an N such that if $n \geq N$, then

$$|a_{j_n} - a| < \epsilon.$$

Therefore, (a_{j_n}) converges to a. This proves that every subsequence of a convergent sequence converges to the same limit.

We now prove the converse. We assume that all subsequences of (a_n) converge to a. But (a_n) is a subsequence of itself (take $j_n = n$ for all $n \geq 1$), thus (a_n) converges to a. This completes the proof of Theorem 4.1.

Example 4.1 For $n \geq 1$, let $a_n = (-1)^n$. Intuitively, it is clear that (a_n) does not converge. We now prove this.

Note that the subsequence (a_{2n}) is the constant sequence 1 and hence converges to 1. On the other hand, (a_{2n+1}) is the constant sequence -1 and hence converges to -1. That is, we have found two subsequences that converge to two distinct limits. By Theorem 4.1 the sequence (a_n) does not converge.

Example 4.2 Let $c > 1$. Define for $n \geq 1$, $a_n = c^{1/n}$. Show that (a_n) converges.

Since $\frac{1}{n+1} < \frac{1}{n}$ and $c > 1$ we have that $a_{n+1} < a_n$ (why?). That is, (a_n) is a decreasing sequence. Moreover, we have for every $n \geq 1$, $c^{1/n} > 1^{1/n} = 1$. Therefore, (a_n) is decreasing and bounded below by 1, it converges.

We are now going to find the limit ℓ of (a_n). Consider the subsequence (a_{2n}). By Theorem 4.1, (a_{2n}) converges to ℓ as well. On the other hand, for every $n \geq 1$,

$$a_{2n} = c^{\frac{1}{2n}} = \sqrt{a}_n.$$

Since (a_n) converges to ℓ, $(\sqrt{a}_n)$ converges to $\sqrt{\ell}$ (why?). Thus, (a_{2n}) converges to ℓ and to $\sqrt{\ell}$. Therefore, $\ell = \sqrt{\ell}$. Either $\ell = 0$ or $\ell = 1$ but ℓ cannot be 0 (why not?). Therefore, $\ell = 1$.

The following result turns out to be useful in a number of situations.

Proposition 4.1 *Assume that the sequence (a_n) does not converge to ℓ. Then, there exists an $\epsilon > 0$ and a subsequence (a_{j_n}) such that for every $n \geq 1$,*

$$|a_{j_n} - \ell| \geq \epsilon.$$

Proof of Proposition 4.1

We do a proof by induction to show the existence of natural numbers

$$j_1 < j_2 < \cdots < j_n < \ldots$$

such that $|a_{j_n} - \ell| \geq \epsilon$ for every $n \geq 1$.

Since the sequence (a_n) does not converge to ℓ, there exists an $\epsilon > 0$ such that for every natural N there is an $n \geq N$ with the property

$$|a_n - \ell| \geq \epsilon.$$

Let $N = 1$, then there exists $n = j_1 \geq 1$ such that $|a_{j_1} - \ell| \geq \epsilon$.

Assume now that there are $j_1 < j_2 < \cdots < j_n$ such that $|a_{j_k} - \ell| \geq \epsilon$ for $k = 1, 2, \ldots, n$. Let $N = j_n + 1$, then there is $j_{n+1} \geq N > j_n$ such that

$$|a_{j_{n+1}} - \ell| \geq \epsilon.$$

This shows the existence of a subsequence of (a_n) with the stated property. This completes the proof of Proposition 4.1.

5 Bolzano-Weierstrass Theorem

As we have seen, a bounded sequence need not converge. However, the following weaker statement holds.

Theorem 5.1 (Bolzano-Weierstrass) *A bounded sequence has at least one convergent subsequence.*

Bolzano-Weierstrass Theorem is one of the fundamental results in analysis. In order to prove this theorem we first show that every sequence (bounded or not) has a monotone subsequence.

Lemma 5.1 *Every sequence has a monotone subsequence.*

Proof of Lemma 5.1
Consider a sequence (a_n). Let

$$A = \{m \in \mathbb{N} : \text{ for all } n > m, a_n \leq a_m\}.$$

There are three possibilities: A may be empty, finite, or infinite.

Assume first that A is empty or finite. We will show that for every $k \geq 2$ there are natural numbers $n_1 < n_2 < \cdots < n_k$ such that

$$a_{n_1} < a_{n_2} < \cdots < a_{n_k}. \tag{5.1}$$

This will give us the existence of a strictly increasing subsequence (a_{n_k}) of (a_n).

If A is empty, let $N = 1$. If A is not empty but finite, let N be the maximum of A (i.e., the largest natural in A). Set $n_1 = N + 1$. Since n_1 is not in A (why not?), there exists at least one natural $n_2 > n_1$ such that $a_{n_2} > a_{n_1}$. This proves (5.1) for $k = 2$.

Assume now that (5.1) holds for some $k \geq 2$. Using that $n_k > N$ we know that n_k is not in A. Hence, there is $n_{k+1} > n_k$ such that

$$a_{n_{k+1}} > a_{n_k}.$$

That is, (5.1) holds for $k+1$. By the Induction Principle (5.1) holds for every $k \geq 2$.

Assume now that A is infinite. Then, there exists a strictly increasing sequence of natural numbers (n_k) in A. For every $k \geq 1$, n_k is in A and $n_{k+1} > n_k$. Hence,

$$a_{n_{k+1}} \leq a_{n_k}.$$

That is, the sequence (a_{n_k}) is decreasing.

This completes the proof of Lemma 5.1.

Proof of Theorem 5.1
Let (a_n) be a bounded sequence. By Lemma 5.1, (a_n) has a monotone subsequence (a_{n_k}). Since (a_n) is bounded so is (a_{n_k}). Hence, (a_{n_k}) is a monotone and bounded sequence. Therefore, it converges. This completes the proof of Theorem 5.1.

Problems

1. Prove that a decreasing sequence converges if and only if it is bounded below.

2. Assume that (a_n) is an increasing sequence, (b_n) is a decreasing sequence, and $a_n \leq b_n$ for all $n \geq 1$.

(a) Show that (a_n) and (b_n) converge.
(b) If in addition to the previous hypotheses we have that $(b_n - a_n)$ converges to 0, prove that (a_n) and (b_n) converge to the same limit.

3. Assume that $|c| < 1$. Show that the sequence (c^n) converges to 0.

4. Let $c > 1$. Show that the sequence (c^n) tends to $+\infty$. That is, show that for every $a > 0$ there exists a natural N such that if $n \geq N$, then $c^n > a$.

5. Let (j_n) be a strictly increasing sequence of natural numbers. Show that for every $n \geq 1$, we have $j_n \geq n$.

6. Let $a > 0$. Show that $(a^{1/n})$ converges to 1.

7. Show that a subsequence of a monotone sequence is also monotone.

8. Suppose that (a_n) is a monotone sequence. Show that (a_n) converges if and only if it has a subsequence which converges.

9. Let $a_n > 0$ for every $n \geq 1$ and assume that

$$\lim_n \frac{a_{n+1}}{a_n} = r < 1.$$

Show that (a_n) converges.

10. Let $a_1 = 1$ and for $n \geq 2$

$$a_n = a_{n-1} + \frac{1}{a_{n-1}}.$$

Show that (a_n) does not converge.

11. We define $r_1 = 1$ and for $n \geq 2$

$$r_n = 1 + \frac{1}{r_{n-1}}.$$

(a) Show that if (r_n) converges, then it must converge to

$$\phi = \frac{1+\sqrt{5}}{2},$$

the so-called golden ratio.

(b) Show that for every $n \geq 2$ we have

$$|r_n - \phi| \leq \frac{1}{\phi}|r_{n-1} - \phi|.$$

(c) Show that (r_n) does converge to ϕ.

12. Consider a sequence (a_n) with the following property. Every subsequence of (a_n) has a subsequence that converges to the same ℓ. Show that (a_n) converges to ℓ.

13. A set C in $\mathbb{R}$ is said to be closed if it has the following property. If $a_n \in C$ for every $n \geq 1$ and if (a_n) converges to ℓ, then ℓ must belong to C.

(a) Give an example of a closed set. Prove your claim.

(b) An open set O in $\mathbb{R}$ is a set whose complement (all the elements not in O) is closed. Show that if O is open, then for every a in O it is possible to find $\epsilon > 0$ such that $(a - \epsilon, a + \epsilon) \subset O$.

6 Cauchy Sequences

The following notion is closely related to convergence and turns out to be quite important. The sequence (a_n) is said to be a **Cauchy sequence** if for every $\epsilon > 0$ there exists a natural N such that if $n \geq N$ and $p \geq N$, then

$$|a_n - a_p| < \epsilon.$$

Proposition 6.1 *A convergent sequence is a Cauchy sequence.*

Proof of Proposition 6.1
Let (a_n) be a sequence converging to ℓ. Let $\epsilon > 0$. There is a natural $N \geq 1$ such that if $n \geq N$, then

$$|a_n - \ell| < \frac{\epsilon}{2}.$$

We also have that if $p \geq N$, then

$$|a_p - \ell| < \frac{\epsilon}{2}.$$

By the triangle inequality

$$|a_n - a_p| = |a_n - \ell + \ell - a_p| \leq |a_n - \ell| + |\ell - a_p| < \frac{\epsilon}{2} + \frac{\epsilon}{2} = \epsilon,$$

for all $n \geq N$ and $p \geq N$. That is, (a_n) is a Cauchy sequence. This completes the proof of Proposition 6.1.

Proposition 6.2 *A Cauchy sequence is a bounded sequence.*

Proof of Proposition 6.2
Let (a_n) be a Cauchy sequence. There is an $N \geq 1$ such that if $n \geq N$ and $p \geq N$, then

$$|a_n - a_p| < 1.$$

We have for all natural numbers n and p

$$|a_n| = |a_n - a_p + a_p| \leq |a_n - a_p| + |a_p|.$$

Let $p = N$, we have for all $n \geq N$

$$|a_n| < 1 + |a_N|.$$

Let m be the maximum of the finite set

$$\{|a_1|, |a_2|, \ldots, |a_{N-1}|\}.$$

Let $m_1 = \max(1 + |a_N|, m)$. Then, $a_n \leq m_1$ for all $n \geq 1$. That is, (a_n) is bounded. This completes the proof of Proposition 6.2.

We now turn to another fundamental property of the real numbers.

Theorem 6.1 *A sequence of real numbers is convergent if and only if it is Cauchy.*

- The definition of convergence involves knowing the limit of the sequence. This is a nontrivial issue. Many times we do not know what the limit is. Thanks to Theorem 6.1 we can now prove convergence of a sequence by proving that the sequence is Cauchy. This does not involve the limit at all!

Proof of Theorem 6.1
We have already proved (Proposition 6.1) that a convergent sequence is Cauchy. We now prove the converse.

Let (a_n) be a Cauchy sequence. By Proposition 6.2, (a_n) must be bounded. By Bolzano-Weierstrass Theorem, (a_n) has a convergent subsequence (a_{n_k}). Denote its limit by ℓ.

For $\epsilon > 0$ there exists a natural N such that if $n \geq N$ and $k \geq N$, then

$$|a_n - a_k| < \frac{\epsilon}{2}.$$

Recall that $n_k \geq k$ for all $k \geq 1$. Hence, if $k \geq N$, then $n_k \geq N$. Therefore, for $k \geq N$ and $n \geq N$ we have

$$|a_n - a_{n_k}| < \frac{\epsilon}{2}.$$

Since (a_{n_k}) converges to ℓ there exists a natural N_1 such that if $k \geq N_1$, then

$$|a_{n_k} - \ell| < \frac{\epsilon}{2}.$$

Let $N_2 = \max(N, N_1)$. For $n \geq N$ and $k \geq N_2$ we have

$$\begin{aligned}
|a_n - \ell| &= |a_n - a_{n_k} + a_{n_k} - \ell| \\
&\leq |a_n - a_{n_k}| + |a_{n_k} - \ell| \\
&< \frac{\epsilon}{2} + \frac{\epsilon}{2} = \epsilon.
\end{aligned}$$

Therefore, for any $\epsilon > 0$ there is a natural N such that if $n \geq N$, then $|a_n - \ell| < \epsilon$. This proves that (a_n) converges. The proof of Theorem 6.1 is complete.

Example 6.1 In this example we show that Theorem 6.1 does not hold for rational numbers. That is, we will give an example of a sequence of rational numbers which is Cauchy but does not converge to a rational number. Let $a_1 = 2$ and for $n \geq 1$

$$a_{n+1} = \frac{1}{2}(a_n + \frac{2}{a_n}).$$

- Our first step is to prove that for every $n \geq 1$, a_n is a strictly positive rational number. We do a proof by induction. Since $a_1 = 2$ the claim is true for $n = 1$. Assume now that a_n is a strictly positive rational. Since

$$a_{n+1} = \frac{1}{2}(a_n + \frac{2}{a_n}),$$

  a_{n+1} is clearly also a strictly positive rational. By the induction Principle this proves our claim.
- We now prove that $a_n > \sqrt{2}$ for every $n \geq 1$. Since $a_1 = 2$ we have $a_1 > \sqrt{2}$. Let $n \geq 1$,

$$a_{n+1}^2 - 2 = \left(\frac{1}{2}(a_n + \frac{2}{a_n})\right)^2 - 2 = \frac{1}{4}(a_n - \frac{2}{a_n})^2 \geq 0.$$

  Since a_n is rational for every n we have in fact the strict inequality

$$\frac{1}{4}(a_n - \frac{2}{a_n})^2 > 0.$$

  Thus, $a_{n+1}^2 - 2 > 0$ and since $a_{n+1} > 0$ we have $a_{n+1} > \sqrt{2}$. In summary, $a_n > \sqrt{2}$ for every $n \geq 1$.
- In this step we show that (a_n) is a strictly decreasing sequence that converges to an irrational number.

$$a_n - a_{n+1} = a_n - \frac{1}{2}(a_n + \frac{2}{a_n}) = \frac{1}{2a_n}(a_n^2 - 2) > 0,$$

  where the last inequality has been proved in the preceding step. This shows that (a_n) is strictly decreasing. We also know that (a_n) is bounded below. Hence, (a_n) converges to a real number ℓ. We now find ℓ.

  Since $a_n > \sqrt{2}$ for every $n \geq 1$, we have $\ell \geq \sqrt{2}$ (why?). Using that

$$a_{n+1} = \frac{1}{2}(a_n + \frac{2}{a_n}),$$

  we have by operations on convergent sequences and the fact that $\ell \neq 0$,

$$\ell = \frac{1}{2}(\ell + \frac{2}{\ell}).$$

Hence, $\ell^2 = 2$ and since $\ell \geq \sqrt{2}$ we have $\ell = \sqrt{2}$.

Since (a_n) is convergent it is Cauchy. Hence, (a_n) is a Cauchy sequence of rational numbers that converges to an irrational number. In the set of rational numbers a Cauchy sequence need not converge!

Problems

1. Assume that (a_n) is a Cauchy sequence of real numbers. Show that (a_n) converges if and only if it has a subsequence that converges.

2. Let $a_n = \sqrt{n}$ and $b_n = a_{n+1} - a_n$ for every $n \geq 1$.

(a) Show that (b_n) converges to 0.
(b) Show that (a_n) is not a Cauchy sequence.

3. Let (a_n) be a sequence with the following property. There is a constant c in $(0, 1)$ such that for every $n \geq 1$

$$|a_{n+1} - a_n| \leq c^n.$$

Show that (a_n) converges.

4. This problem is a generalization of Example 6.1. Let $c > 0$ and let a_1 be a number strictly larger than $\sqrt{c}$. For $n \geq 1$, let

$$a_{n+1} = \frac{1}{2}(a_n + \frac{c}{a_n}).$$

(a) Show that for every $n \geq 1$, $a_n > \sqrt{c}$.
(b) Show that (a_n) is a strictly decreasing sequence.
(c) Show that (a_n) converges and find its limit.

5. Show that there exists a strictly increasing sequence (a_n) of natural numbers such that the sequence $(\sin(a_n))$ converges.

6. Let a_1 and a_2 be two real numbers and for $n \geq 2$ let

$$a_{n+1} = \frac{1}{2}(a_n + a_{n-1}).$$

(a) Show that

$$|a_{n+1} - a_n| = \frac{1}{2}|a_n - a_{n-1}|.$$

(b) Show that (a_n) converges.

7 The Fundamental Property Revisited

We introduced the real numbers as a set which contains the rational numbers and has the Fundamental Property (i.e., every nonempty set of reals which is bounded above has a least upper bound). We have used the Fundamental Property to prove important properties of the reals. We will now see that several of these properties are equivalent.

Theorem 7.1 *The following properties are equivalent on the real numbers.*

(i) *Every nonempty set of real numbers which is bounded above has a least upper bound.*
(ii) *An increasing sequence of real numbers which is bounded above converges.*
(iii) *Every bounded sequence of real numbers has a convergent subsequence.*
(iv) *Every Cauchy sequence of real numbers converges.*

Theorem 7.1 shows that the Fundamental Property of $\mathbb{R}$ is equivalent to three other (fundamental) properties. None of these properties hold for the set of rational numbers. The set of real numbers is the smallest set containing the rational numbers and having these four equivalent properties.

Proof of Theorem 7.1
The implication (i) $\Rightarrow$ (ii) is proved in Theorem 3.1. The implication (ii) $\Rightarrow$ (iii) is proved in Theorem 5.1. The implication (iii) $\Rightarrow$ (iv) is proved in Theorem 6.1. The only implication to be proved is therefore (iv) $\Rightarrow$ (i). We prove this now.

Assume that A is a nonempty subset of $\mathbb{R}$ which is bounded above. We have to show that A has a least upper bound. We are going to show the existence of sequences (a_n) and (b_n) with the following properties.

(1) For any $n \geq 1$, a_n is in A. Moreover, (a_n) is an increasing sequence.
(2) For any $n \geq 1$, b_n is an upper bound of A. Furthermore, (b_n) is a decreasing sequence.
(3) For any $n \geq 1, 0 \leq b_n - a_n \leq \frac{b_1 - a_1}{2^{n-1}}$.

Since $A \neq \emptyset$ there is an a_1 in A and since A is bounded above there is a b_1 in $\mathbb{R}$ which is an upper bound of A. Hence, the first three properties hold for $n = 1$.

Let $n \geq 1$. Assume the existence of $a_1 \leq a_2, \cdots \leq a_n$ in A and $b_1 \geq b_2 \geq \cdots \geq b_n$ for which (1), (2), and (3) hold. Let

$$c_n = \frac{a_n + b_n}{2}.$$

There are two possibilities.

- If c_n is an upper bound of A, we set

$$a_{n+1} = a_n \text{ and } b_{n+1} = c_n.$$

Clearly, a_{n+1} belongs to A and b_{n+1} is an upper bound of A. Moreover,

$$0 \leq b_{n+1} - a_{n+1} = \frac{a_n + b_n}{2} - a_n = \frac{b_n - a_n}{2} \leq \frac{b_1 - a_1}{2^n},$$

where the last inequality comes from the induction hypothesis. Hence, the three properties hold for $n + 1$.

- If c_n is not an upper bound of A, there exists an a_{n+1} in A such that $a_{n+1} > c_n$ (why?). We set $b_{n+1} = b_n$. Clearly, a_{n+1} belongs to A and b_{n+1} is an upper bound of A. Moreover,

$$0 \leq b_{n+1} - a_{n+1} < b_n - c_n = \frac{b_n - a_n}{2} \leq \frac{b_1 - a_1}{2^n}.$$

Hence, property (3) holds for $n + 1$.

Note that in both cases we have $a_{n+1} \geq a_n$ and $b_{n+1} \leq b_n$. This completes the proof by induction that there are sequences (a_n) and (b_n) with properties (1) through (3).

To finish the proof of Theorem 7.1 we will show that the sequence (b_n) converges to some ℓ and that ℓ is the least upper bound of A.

We first show that (b_n) is Cauchy. By our definition of (b_n), for each $n \geq 1$, either $b_n - b_{n+1} = 0$ or

$$b_n - b_{n+1} = b_n - c_n = \frac{b_n - a_n}{2}.$$

In both cases we have for all $n \geq 1$

$$0 \leq b_n - b_{n+1} \leq \frac{b_n - a_n}{2}.$$

Hence, by Property (3) we have for all $n \geq 1$

$$0 \leq b_n - b_{n+1} \leq \frac{b_1 - a_1}{2^n}. \tag{7.1}$$

Let $p > n$. We have,

$$0 \leq b_n - b_p = b_n - b_{n+1} + b_{n+1} - b_{n+2} + \cdots + b_{p-1} - b_p.$$

Hence, by (7.1)

$$0 \leq b_n - b_p \leq \frac{b_1 - a_1}{2^n} + \frac{b_1 - a_1}{2^{n+1}} + \cdots + \frac{b_1 - a_1}{2^{p-1}}.$$

Therefore,

$$0 \leq b_n - b_p \leq \frac{b_1 - a_1}{2^n} \sum_{k=0}^{p-1} \frac{1}{2^k} < \frac{b_1 - a_1}{2^n} \sum_{k=0}^{\infty} \frac{1}{2^k} = \frac{b_1 - a_1}{2^{n-1}},$$

where the last equality uses the sum of a geometric series. For any $\epsilon > 0$ we can find N so that if $n \geq N$, then

$$\frac{b_1 - a_1}{2^{n-1}} < \epsilon.$$

Hence, for all $p > n \geq N$ we have $0 \leq b_n - b_p < \epsilon$. That is, the sequence (b_n) is Cauchy and therefore converges to some real number ℓ.

We now show that ℓ is an upper bound of A. For every a in A and $n \geq 1$ we have by property (2) that $a \leq b_n$. Letting n go to infinity gives $a \leq \ell$. Since this is true for every a in A, ℓ is an upper bound of A.

Finally, we need to show that ℓ is the least upper bound of A. By property (3) the sequence $(b_n - a_n)$ converges to 0 (why?). Hence, (a_n) converges to ℓ as well. If we take any $b < \ell$ there is an N such that if $n \geq N$, then

$$b < a_n \leq \ell.$$

Since the sequence (a_n) is in A, this shows that b cannot be an upper bound of A. Therefore, ℓ is the least upper bound of A. This concludes the proof of Theorem 7.1.

Chapter 3
Limits Superior and Inferior of a Sequence

1 Infinite Limits

A sequence (a_n) is said to **tend to positive infinity** if for any real m there exists a natural N such that if $n \geq N$, then $a_n > m$. The notation will be

$$\lim_n a_n = +\infty.$$

A sequence (a_n) is said to **tend to negative infinity** if for any real m there exists a natural N such that if $n \geq N$, then $a_n < m$. The notation will be

$$\lim_n a_n = -\infty.$$

We will say that the sequence **converges** if the sequence has a finite limit. When we say that the limit of a sequence **exists** the limit may be finite or infinite.

Proposition 1.1 *Assume (a_n) is increasing and not bounded above. Then,*

$$\lim_n a_n = +\infty.$$

There is, of course, a symmetric result for decreasing sequences. See the Problems.

Proof of Proposition 1.1
Let m be a real number. Since (a_n) is not bounded above there exists a natural N such that $a_N > m$. Since (a_n) is increasing, we have for all $n \geq N$

$$a_n \geq a_N > m.$$

Hence, $\lim_n a_n = +\infty$. This completes the proof of Proposition 1.1.

R. B. Schinazi, *From Classical to Modern Analysis*,
https://doi.org/10.1007/978-3-319-94583-5_3

Proposition 1.2 *Assume (a_n) is a sequence of real numbers. Then, (a_n) tends to positive infinity if and only if every subsequence of (a_n) tends to positive infinity.*

Note that Proposition 1.2 is analogous to Theorem 4.1 in Chapter 2 for finite limits.

Proof of Proposition 1.2
Assume that (a_n) tends to positive infinity. Let (a_{j_n}) be a subsequence of (a_n) and m be a real number. There is a natural N such that if $n \geq N$, then $a_n > m$. Since $j_k \geq k$ for every natural k, if $n \geq N$, then $j_n \geq N$ and

$$a_{j_n} > m,$$

for all $n \geq N$. That is, $\lim_n a_{j_n} = +\infty$.

The converse is trivial. If every subsequence of (a_n) tends to positive infinity so does the whole sequence (a_n) since (a_n) is a subsequence of itself. This completes the proof of Proposition 1.2.

2 Limit Superior and Limit Inferior

The following result establishes the existence of a limit superior.

Proposition 2.1 *Let (a_n) be a sequence of real numbers.*

- *If (a_n) is bounded above, then for $n \geq 1$ let*

$$b_n = \sup\{a_n, a_{n+1}, \ldots\}.$$

Then, (b_n) is a decreasing sequence. Hence, $\lim_n b_n$ exists (it may be finite or $-\infty$). We define the **limit superior** *of (a_n) as*

$$\limsup_n a_n = \lim_n b_n.$$

- *If (a_n) is not bounded above we set*

$$\limsup_n a_n = +\infty.$$

Note that $\limsup_n a_n = +\infty$ if and only if (a_n) is not bounded above.

Proof of Proposition 2.1
Assume the sequence (a_n) is bounded above by some real number m. Then, for each $n \geq 1$, the set

$$B_n = \{a_n, a_{n+1}, \ldots\}$$

is also bounded above by m (why?). Hence, the set B_n is not empty and is bounded above. Therefore, B_n has a least upper bound that we denote by b_n. Note that for any $n \geq 1$, $B_{n+1} \subset B_n$. Hence, $b_{n+1} \leq b_n$ (why?) for every $n \geq 1$. Therefore, (b_n) is a decreasing sequence. There are two possibilities. Either (b_n) is bounded below and then it converges to a finite limit. Or, (b_n) is not bounded below and we have $\lim_n b_n = -\infty$. In both cases $\lim_n b_n$ exists and we can define

$$\limsup_n a_n = \lim_n b_n.$$

This completes the proof of Proposition 2.1.

We now state an analogous result for the limit inferior of a sequence.

Proposition 2.2 *Let (a_n) be a sequence of real numbers.*
- *If (a_n) is bounded below, then for $n \geq 1$ let*

$$c_n = \inf\{a_n, a_{n+1}, \dots\}.$$

This defines an increasing sequence (c_n) of real numbers. Hence, $\lim_n c_n$ *exists (it may be finite or $+\infty$). We define the* **limit inferior** *of (a_n) as*

$$\liminf_n a_n = \lim_n c_n.$$

- *If (a_n) is not bounded below, then we set*

$$\liminf_n a_n = -\infty.$$

Note that $\liminf_n a_n = -\infty$ if and only if (a_n) is not bounded below.

The proof of Proposition 2.2 is quite similar to the proof of Proposition 2.1 and is left to the reader.

- In contrast to limits which need not exist lim sup and lim inf always exist. As we will see in many applications lim sup and lim inf are quite useful in determining whether a sequence has a limit.

Example 2.1 For $n \geq 1$ let $a_n = (-1)^n$. What are the $\limsup_n a_n$ and $\liminf_n a_n$?

First note that for every $n \geq 1$ we have

$$\{a_n, a_{n+1}, \dots\} = \{-1, 1\}.$$

Hence, $\inf\{a_n, a_{n+1}, \dots\} = -1$ and $\sup\{a_n, a_{n+1}, \dots\} = 1$ for every $n \geq 1$. Therefore, $\limsup_n a_n = 1$ and $\liminf_n a_n = -1$.

Example 2.2 For $n \geq 1$ let $a_n = n$. What are the $\limsup_n a_n$ and $\liminf_n a_n$?

Since (a_n) is not bounded above we set $\limsup_n a_n = +\infty$.

Note that for every $n \geq 1$ we have

$$\{a_n, a_{n+1}, \dots\} = \{n, n+1, \dots\}.$$

Hence, $\inf\{a_n, a_{n+1}, \dots\} = n$ for every $n \geq 1$. Therefore, $\liminf_n a_n = +\infty$.

Proposition 2.3 *Let (a_n) be a sequence of real numbers.*

- *There exists a subsequence of (a_n) whose limit is* $\limsup_n a_n$.
- *There exists a subsequence of (a_n) whose limit is* $\liminf_n a_n$.

Proof of Proposition 2.3

We prove the statement for $\limsup_n a_n$. The proof for $\liminf_n a_n$ is analogous and we omit it.

There are three cases to consider.

- Assume first that $\limsup_n a_n$ is a real number $\bar{\ell}$.

For $n \geq 1$, let

$$B_n = \{a_n, a_{n+1}, \dots\}$$

and

$$b_n = \sup B_n.$$

We will prove by induction that for every $n \geq 1$ there exist natural numbers $j_1 < j_2 < \dots < j_n$ such that

$$0 \leq b_{k_n} - a_{j_n} < \frac{1}{2^n}, \tag{2.1}$$

where $k_1 = 1$ and for $n \geq 2$, $k_n = j_{n-1} + 1$.

Since b_1 is the least upper bound of B_1, $b_1 - \frac{1}{2}$ is not an upper bound of B_1. Hence, there is a_{j_1} in B_1 such that $a_{j_1} > b_1 - \frac{1}{2}$, where $j_1 \geq 1$ is a natural number. Therefore, letting $k_1 = 1$ we get

$$0 \leq b_{k_1} - a_{j_1} < \frac{1}{2}.$$

This shows that (2.1) holds for $n = 1$.

Assume now that (2.1) holds for n. Let $k_{n+1} = j_n + 1$. We have that

$$b_{k_{n+1}} - \frac{1}{2^{n+1}}$$

is not an upper bound of $B_{k_{n+1}}$ (why?). Hence, there exists a natural $j_{n+1} \geq k_{n+1} = j_n + 1$ such that

$$0 \leq b_{k_{n+1}} - a_{j_{n+1}} < \frac{1}{2^{n+1}}.$$

Therefore, (2.1) holds for $n + 1$. By induction, (2.1) holds for every $n \geq 1$.

Since (b_{k_n}) is a subsequence of (b_n) and (b_n) converges to $\bar{\ell}$, (b_{k_n}) converges to $\bar{\ell}$ as well. Letting n go to infinity in (2.1) we see that the sequence $(b_{k_n} - a_{j_n})$ converges to 0. Therefore, (a_{j_n}) converges to $\bar{\ell} = \limsup_n a_n$. That is, there exists a subsequence of (a_n) whose limit is $\limsup_n a_n$.

- We now turn to the case $\limsup_n a_n = +\infty$. We will prove that for every natural k there exist natural numbers $n_1 < n_2 < \cdots < n_k$ such that

$$a_{n_j} > j \text{ for } j = 1, 2, \ldots, k. \tag{2.2}$$

Since $\limsup_n a_n = +\infty$, (a_n) is not bounded above. In particular, there is a natural n_1 such that $a_{n_1} > 1$. Hence, (2.2) holds for $k = 1$. Assume now that (2.2) holds for $k \geq 1$. By contradiction, assume (2.2) does not hold for $k + 1$. That is, there is no $n > n_k$ such that $a_n > k + 1$. Hence, $a_n \leq k + 1$ for all $n > n_k$. Since

$$\{a_1, a_2, \ldots, a_{n_k}\}$$

is a finite set it has a largest element m_1. Let $m = \max(m_1, k + 1)$, then $a_n \leq m$ for every $n \geq 1$ (why?). That is, (a_n) is bounded above. We have a contradiction. Therefore, there exists $n_{k+1} > n_k$ such that

$$a_{n_{k+1}} > k + 1.$$

By induction, this shows the existence of a subsequence (a_{n_k}) with the property $a_{n_k} > k$ for every $k \geq 1$. It is now easy to show that (a_{n_k}) tends to $+\infty$ which is the $\limsup_n a_n$.

- The third and final case is $\limsup_n a_n = -\infty$. The sequence (a_n) must be bounded above. Let

$$b_n = \sup\{a_n, a_{n+1}, \ldots\}$$

for every $n \geq 1$. We have $a_n \leq b_n$ for every $n \geq 1$ (why?). Since (b_n) tends to $-\infty$ so does (a_n). Hence, $\lim_n a_n = \limsup_n a_n$.

This completes the proof of Proposition 2.3.

Proposition 2.4 *Let (a_n) be a sequence of real numbers. We have*

$$\liminf_n a_n \leq \limsup_n a_n.$$

- The inequality in Proposition 2.4 may involve $+\infty$ or $-\infty$. Rules involving infinity are the following. For any real number a we have

$$-\infty < a < +\infty.$$

For any ℓ (finite or infinite) we have

$$-\infty \leq \ell \leq +\infty.$$

Proof of Proposition 2.4
If $\liminf_n a_n = -\infty$ there is nothing to prove since we always have $\limsup_n a_n \geq -\infty$. Assume now that $\liminf_n a_n$ is not $-\infty$. Then, (a_n) is bounded below and the sequence

$$c_n = \inf\{a_n, a_{n+1}, \ldots\}$$

can be defined. We have $\lim_n c_n = \liminf_n a_n$.

On the other hand, by Proposition 2.3 there is a subsequence (a_{j_n}) that converges to $\limsup_n a_n$. Note that for every $n \geq 1$,

$$a_{j_n} \geq c_{j_n}.$$

Since (c_{j_n}) is a subsequence of (c_n), (c_{j_n}) converges to $\liminf_n a_n$. Hence,

$$\lim_n a_{j_n} = \limsup_n a_n \geq \lim_n c_{j_n} = \liminf_n a_n.$$

This completes the proof of Proposition 2.4.

The following result explains the names limit superior and limit inferior.

Proposition 2.5 *Let (a_n) be a sequence of real numbers. Assume that a subsequence (a_{j_n}) has a limit. Then,*

$$\liminf_n a_n \leq \lim_n a_{j_n} \leq \limsup_n a_n.$$

In words, if a subsequence has a limit it must be between the limits inferior and superior.

The proof of Proposition 2.5 can be obtained with the same ideas used in the proof of Proposition 2.4. We leave this proof to the reader.

Theorem 2.1 *Let (a_n) be a sequence of real numbers. The limit of (a_n) exists if and only if*

$$\liminf_n a_n = \limsup_n a_n.$$

If the equality holds, then

$$\lim_n a_n = \liminf_n a_n = \limsup_n a_n.$$

Proof of Theorem 2.1
Assume that $\liminf_n a_n \neq \limsup_n a_n$. Then, by Proposition 2.3 we have two subsequences, one having limit $\liminf_n a_n$ and the other one having limit $\limsup_n a_n$. But we know (a_n) has a limit if and only every subsequence has the same limit. This was proved for a finite limit in Theorem 4.1 in Chapter 2 and for an infinite limit in Proposition 1.2. Thus, if $\liminf_n a_n \neq \limsup_n a_n$, then the limit of (a_n) does not exist. This proves one implication.

For the converse, assume that $\liminf_n a_n = \limsup_n a_n$. Consider the following cases.

- Assume $\liminf_n a_n = \limsup_n a_n = \ell$ where ℓ is a real number. By contradiction, assume that (a_n) does not converge to ℓ. Then, there is an $\epsilon > 0$ and a subsequence (a_{j_n}) such that for all $n \geq 1$

$$|a_{j_n} - \ell| \geq \epsilon. \tag{2.3}$$

  Since $\liminf_n a_n$ and $\limsup_n a_n$ are both finite, the sequence (a_n) is bounded below and above. Hence, the subsequence (a_{j_n}) is bounded as well. By Bolzano-Weierstrass Theorem, there exists a subsequence of (a_{j_n}) that we denote by $(a_{j'_n})$ that converges. By Proposition 2.5

$$\liminf_n a_n \leq \lim_n a_{j'_n} \leq \limsup_n a_n.$$

  Since $\liminf_n a_n = \limsup_n a_n = \ell$ the limit of $(a_{j'_n})$ must be ℓ. On the other hand, by (2.3) we have for every $n \geq 1$ that

$$|a_{j'_n} - \ell| \geq \epsilon.$$

  Hence, $(a_{j'_n})$ converges to ℓ and does not converge to ℓ. We have a contradiction. The sequence (a_n) converges to ℓ.
- Assume $\liminf_n a_n = \limsup_n a_n = +\infty$. We must have that (a_n) is bounded below (why?). So we can define for $n \geq 1$,

$$c_n = \inf\{a_n, a_{n+1}, \dots\}.$$

  For every $n \geq 1$, $c_n \leq a_n$. Since $\lim_n c_n = +\infty$ we have $\lim_n a_n = +\infty$.
- The last case to consider is $\liminf_n a_n = \limsup_n a_n = -\infty$. It is treated similarly to the preceding case and we omit it.

  This completes the proof of Theorem 2.1.

Proposition 2.6 *Assume that* $\liminf_n a_n = \ell$ *is finite. For any* $b < \ell$ *there exists a natural* N *such that if* $n \geq N$*, then*

$$a_n > b.$$

There is a similar result for the limit superior. See the problems.

Proof of Proposition 2.6
By contradiction, assume that there is a $b < \ell$ such that for every natural N there exists an $n \geq N$ with $a_n \leq b$. We will show the existence of a subsequence that converges to a limit smaller than b. This will contradict Proposition 2.5.

Let $N = 1$, there exists $j_1 \geq 1$ such that $a_{j_1} \leq b$. Assume that there are natural numbers $j_1 < j_2 < \cdots < j_k$ such that

$$a_{j_p} \leq b \text{ for } p = 1, \ldots, k.$$

Let $N = j_k + 1 > j_k$, there exist a $j_{k+1} \geq N$ such that

$$a_{j_{k+1}} \leq b.$$

By induction, this proves the existence of a subsequence (a_{j_k}) with the property, $a_{j_k} \leq b$ for every $k \geq 1$. Since $\liminf_n a_n$ is finite we know that (a_n) is bounded below. Hence, so is the subsequence (a_{j_k}). We also know that (a_{j_k}) is bounded above by b. Since (a_{j_k}) is bounded there is a subsequence $(a_{j'_k})$ of (a_{j_k}) that converges (why?). The limit of $(a_{j'_k})$ is smaller or equal to b (why?). Since $b < \liminf_n a_n$, this contradicts Proposition 2.5. Hence, there exists a natural N such that if $n \geq N$, then $a_n > b$. The proof of Proposition 2.6 is complete.

3 Operations on Limits

The following result establishes a helpful relation between limit inferior and limit superior.

Proposition 3.1 *Let* (a_n) *be a sequence of real numbers. We have*

$$\liminf_n a_n = -\limsup_n (-a_n).$$

Proof of Proposition 3.1
As always there are three cases to consider.

- Assume that $\liminf_n a_n = -\infty$. Then (a_n) is not bounded below. Therefore, $(-a_n)$ is not bounded above and

$$\limsup_n(-a_n) = +\infty = -\liminf_n a_n.$$

- Assume that $\liminf_n a_n = +\infty$. Then, we must have $\limsup_n a_n = +\infty$ (why?) and $\lim_n a_n = +\infty$. This implies, $\lim_n(-a_n) = -\infty$. Therefore,

$$\limsup_n(-a_n) = \lim_n(-a_n) = -\infty = -\liminf_n a_n.$$

- Assume that $\liminf_n a_n$ is a real number. Then, (a_n) is bounded below and we can define for every $n \geq 1$

$$c_n = \inf\{a_n, a_{n+1}, \dots\}.$$

The sequence $(-a_n)$ is bounded above. We define for every $n \geq 1$

$$b_n = \sup\{-a_n, -a_{n+1}, \dots\}.$$

We have for all $k \geq n$, $a_k \geq c_n$. Therefore, $-a_k \leq -c_n$ for $k \geq n$. That is, $-c_n$ is an upper bound of

$$\{-a_n, -a_{n+1}, \dots\}.$$

Since b_n is the *least* upper bound of this set, we have for all $n \geq 1$

$$b_n \leq -c_n. \tag{3.1}$$

We now prove the reverse inequality. We have for all $k \geq n$, $-a_k \leq b_n$. Therefore, $a_k \geq -b_n$ for $k \geq n$. That is, $-b_n$ is a lower bound of

$$\{a_n, a_{n+1}, \dots\}.$$

Since c_n is the *greatest* lower bound of this set, we have for all $n \geq 1$

$$-b_n \leq c_n. \tag{3.2}$$

By (3.1) and (3.2) we get that $c_n = -b_n$ for every $n \geq 1$. Therefore,

$$\lim_n c_n = \liminf a_n = -\lim_n b_n = -\limsup_n(-a_n).$$

This completes the proof of Proposition 3.1.

The following easy result is quite useful.

Proposition 3.2 *Let (r_n) and (s_n) be two sequences such that $r_n \leq s_n$ for all $n \geq N$, where N is a fixed natural number. Then,*

(i) $\limsup_n r_n \leq \limsup_n s_n$.
(ii) $\liminf_n r_n \leq \liminf_n s_n$.

Proof of Proposition 3.2
We prove (i) and leave (ii) to the reader.

If $\limsup_n s_n = +\infty$ there is nothing to prove. Assume that $\limsup_n s_n < +\infty$. Define for every $n \geq 1$

$$b_n = \sup\{s_n, s_{n+1}, \ldots, \}.$$

We have for $k \geq n \geq N$ that

$$r_k \leq s_k \leq b_n.$$

Hence, for $n \geq N$, b_n is an upper bound of

$$\{r_n, r_{n+1}, \ldots, \}.$$

Let $d_n = \sup\{r_n, r_{n+1}, \ldots, \}$ for $n \geq N$. Then, $d_n \leq b_n$ for every $n \geq N$. Therefore,

$$\lim_n d_n \leq \lim_n b_n.$$

Since, $\lim_n d_n = \limsup_n r_n$ and $\lim_n b_n = \limsup_n s_n$, the proof of Proposition 3.2 is complete.

Proposition 3.3 *For any sequences (a_n) and (b_n) we have*

$$\liminf_n (a_n + b_n) \geq \liminf_n a_n + \liminf_n b_n.$$

- The inequality above involves sums of possible infinite terms. Here are our rules for such sums.

$$+\infty + \infty = +\infty.$$

$$-\infty - \infty = -\infty.$$

If c is a real number

$$c + \infty = +\infty.$$

$$c - \infty = -\infty.$$

The sums $+\infty - \infty$ and $-\infty + \infty$ are not defined. The inequality in Proposition 3.3 is true provided the r.h.s. is defined. Note that the l.h.s. is always defined.

Example 3.1 Let $a_n = (-1)^n$ and $b_n = (-1)^{n+1}$ for all $n \geq 1$. We have

$$\liminf_n a_n = \liminf_n b_n = -1.$$

On the other hand, $a_n + b_n = 0$ for every $n \geq 1$ and therefore

$$\liminf_n (a_n + b_n) = 0.$$

This provides an example of strict inequality in Proposition 3.3.

Proof of Proposition 3.3

- Assume that $\liminf_n a_n$ and $\liminf_n b_n$ are both finite. Let $\epsilon > 0$. By Proposition 2.6 there is a natural N_1 such that for $n \geq N_1$ then

$$a_n \geq \liminf_n a_n - \frac{\epsilon}{2}.$$

Similarly, there is a natural N_2 such that for $n \geq N_2$ then

$$b_n \geq \liminf_n b_n - \frac{\epsilon}{2}.$$

For $n \geq N = \max(N_1, N_2)$ we have

$$a_n + b_n \geq \liminf_n a_n + \liminf_n b_n - \epsilon.$$

By Proposition 3.2, the preceding inequality implies

$$\liminf_n (a_n + b_n) \geq \liminf_n a_n + \liminf_n b_n - \epsilon.$$

Since this is true for every $\epsilon > 0$ we have

$$\liminf_n (a_n + b_n) \geq \liminf_n a_n + \liminf_n b_n.$$

- Assume that $\liminf_n a_n = -\infty$. Then, we do not allow $\liminf_n b_n = +\infty$. Hence,

$$\liminf_n a_n + \liminf_n b_n = -\infty,$$

and the inequality

$$\liminf_n (a_n + b_n) \geq \liminf_n a_n + \liminf_n b_n$$

holds.

- Assume that $\liminf_n a_n = +\infty$. Since $\liminf_n a_n \le \limsup_n a_n$ we have

$$\lim_n a_n = \liminf_n a_n = \limsup_n a_n = +\infty.$$

Since $\liminf_n a_n = +\infty$ we do not allow $\liminf_n b_n = -\infty$. Therefore, (b_n) must be bounded below by some m.

Hence, for every $n \ge 1$

$$a_n + b_n \ge a_n + m.$$

Since the r.h.s. tends to $+\infty$, so does the l.h.s. In particular,

$$\liminf_n (a_n + b_n) = +\infty.$$

Since $\liminf_n a_n = +\infty$ and $\liminf_n b_n \ne -\infty$,

$$\liminf_n a_n + \liminf_n b_n = +\infty.$$

Therefore,

$$\liminf_n (a_n + b_n) = \liminf_n a_n + \liminf_n b_n.$$

Both sides are actually equal to $+\infty$. This completes the proof of Proposition 3.3.

Proposition 3.4 *If (a_n) converges to a finite limit and (b_n) is any sequence,*

$$\liminf_n (a_n + b_n) = \lim_n a_n + \liminf_n b_n.$$

Proof of Proposition 3.4
Assume that (a_n) converges to some real ℓ. We know that (b_n) has a subsequence (b_{j_n}) such that

$$\lim_n b_{j_n} = \liminf_n b_n.$$

Hence,

$$\lim_n (a_{j_n} + b_{j_n}) = \ell + \liminf_n b_n.$$

Note that since ℓ is finite the r.h.s. is defined even if $\liminf_n b_n$ is infinite. By Proposition 2.5 we have

$$\liminf_n (a_n + b_n) \le \lim_n (a_{j_n} + b_{j_n}).$$

Thus,

$$\liminf_n (a_n + b_n) \leq \ell + \liminf_n b_n.$$

But, by Proposition 3.3 we have

$$\liminf_n (a_n + b_n) \geq \liminf_n a_n + \liminf_n b_n.$$

Since

$$\liminf_n a_n + \liminf_n b_n = \ell + \liminf_n b_n,$$

we have

$$\liminf_n (a_n + b_n) = \lim_n a_n + \liminf_n b_n.$$

This completes the proof of Proposition 3.4.

Proposition 3.5 *For any sequences (a_n) and (b_n) we have*

$$\limsup_n (a_n + b_n) \leq \limsup_n a_n + \limsup_n b_n,$$

provided the r.h.s. is defined.

Proof of Proposition 3.5
By Proposition 3.3 we have

$$\liminf_n (-a_n - b_n) \geq \liminf_n (-a_n) + \liminf_n (-b_n).$$

Using Proposition 3.1

$$-\limsup_n (a_n + b_n) \geq -\limsup_n (a_n) - \limsup_n (b_n),$$

and therefore

$$\limsup_n (a_n + b_n) \leq \limsup_n a_n + \limsup_n b_n.$$

This completes the proof of Proposition 3.5.

Problems

1. Show that if (a_n) is decreasing but not bounded below it tends to $-\infty$.

2.

(a) Give an example of a sequence which is not bounded above and does not tend to $+\infty$.
(b) Show that if a sequence is not bounded above it must have a subsequence that goes to $+\infty$.

3. Assume that $a_n \geq b_n$ for every $n \geq 1$. Assume also that (b_n) tends to $+\infty$. Show that (a_n) tends to $+\infty$ as well.

4. Show that the sequence (a_n) tends to $-\infty$ if and only if every subsequence of (a_n) tends to $-\infty$.

5. Give an example of a sequence (a_n) for which $\liminf_n a_n = -\infty$ and $\limsup_n a_n = +\infty$.

6. Prove Proposition 2.2.

7. Assume that (a_n) converges to a finite limit and (b_n) is any sequence. Show that

$$\limsup_n (a_n + b_n) = \lim_n a_n + \limsup_n b_n.$$

8. Prove Proposition 2.5.

9. Assume that there exists a real m and a natural N such that $a_n \geq m$ for $n \geq N$. Show that

$$\liminf_n a_n \geq m.$$

10.

(a) Show that if $\limsup_n a_n = -\infty$, then $\lim_n a_n = -\infty$.
(b) Show that if $\liminf_n a_n = +\infty$, then $\lim_n a_n = +\infty$.

11. Let (a_n) and (b_n) be two sequences such that

$$\liminf_n a_n \leq \liminf_n b_n \leq \limsup_n b_n \leq \limsup_n a_n.$$

Show that if (a_n) has a limit so does (b_n).

12. Let (a_{j_n}) be a subsequence of (a_n). Show that

$$\liminf_n a_n \leq \liminf_n a_{j_n} \leq \limsup_n a_{j_n} \leq \limsup_n a_n.$$

13. State and prove a result for the limit inferior which is analogous to Proposition 2.6.

14. Prove Proposition 3.2 (ii).

15. Let (a_n) be a sequence and for $n \geq 1$ let

$$b_n = \frac{1}{n}(a_1 + \cdots + a_n).$$

Assume that $\limsup_n a_n$ is a real number $\bar{\ell}$.

(a) Show that for every $\epsilon > 0$ there exists a natural N such that for $n \geq N$ we have

$$a_n < \bar{\ell} + \epsilon.$$

(b) Use (a) to show that for every $\epsilon > 0$

$$\limsup_n b_n \leq \bar{\ell} + \epsilon.$$

(c) Show that

$$\limsup_n b_n \leq \limsup_n a_n.$$

(d) By using steps analogous to (a), (b), and (c) show that

$$\liminf_n a_n \leq \liminf_n b_n.$$

(e) Show that if (a_n) has a limit so does (b_n).

16. Use the notation of Problem 15.. In Problem 15. we proved that

$$\limsup_n b_n \leq \limsup_n a_n$$

when $\limsup_n a_n$ is finite. Show that the inequality remains true if $\limsup_n a_n$ is infinite.

17. Let (a_n) and (b_n) be sequence of positive numbers.

(a) Show that

$$\limsup_n (a_n b_n) \leq \limsup_n a_n \limsup_n b_n.$$

(b) Give an example for which the inequality in (a) is strict.

Chapter 4
Numerical Series

1 First Properties

Let (a_n) be a sequence of real numbers. Define the sequence (s_n) by

$$s_n = a_1 + a_2 + \cdots + a_n = \sum_{k=1}^{n} a_k,$$

for every $n \geq 1$. The sequence (s_n) is said to be the sequence of partial sums. **If** $\lim_n s_n$ exists, then we denote the limit of (s_n) by

$$\lim_n s_n = \sum_{n=1}^{\infty} a_n.$$

The limit

$$\sum_{n=1}^{\infty} a_n$$

is called a **series**.

Example 1.1 Let r be a real number and $a_n = r^n$ for every $n \geq 1$. This geometric sequence will yield the **geometric series**.

For $n \geq 0$, let

$$s_n = \sum_{k=0}^{n} r^k.$$

R. B. Schinazi, *From Classical to Modern Analysis*,
https://doi.org/10.1007/978-3-319-94583-5_4

If $r = 1$, then $s_n = n + 1$ for every $n \geq 0$ and therefore (s_n) tends to $+\infty$. We set

$$\sum_{n=0}^{\infty} a_n = +\infty.$$

For $r \neq 1$ we have for all $n \geq 0$

$$s_n = \frac{r^{n+1} - 1}{r - 1}.$$

There are three different cases.

- If $|r| < 1$, then $\lim_n r^{n+1} = 0$ and

$$\sum_{n=0}^{\infty} r^n = \lim_n s_n = \frac{1}{1-r}.$$

- If $r > 1$, then $\lim_n r^{n+1} = +\infty$ and

$$\sum_{n=0}^{\infty} r^n = \lim_n s_n = +\infty.$$

- If $r \leq -1$, then $\lim_n r^{n+1}$ does not exist (why?) and the geometric series is said to diverge.

In summary, the geometric series exhibits three different behaviors. If $|r| < 1$ it converges, if $r \geq 1$ it does not converge (to a finite limit) but it tends to infinity so we can write

$$\sum_{n=0}^{\infty} r^n = +\infty.$$

Finally, if $r \leq -1$ the series diverges and the symbol $\sum_{n=0}^{\infty} r^n$ is not defined.

Proposition 1.1 (Divergence Test) *If the series $\sum_{k=1}^{\infty} a_k$ converges to a finite limit, then the sequence (a_n) converges to 0.*

This test can show divergence of the series (if (a_n) does not converge to 0) but it can never show convergence of the series.

Proof of Proposition 1.1
Let

$$s_n = \sum_{k=1}^{n} a_k,$$

for every $n \geq 1$.

Assume that the series $\sum_{k=1}^{\infty} a_k$ converges. This means that the sequence of partial sums (s_n) converges to some ℓ. The sequence (s_{n-1}) converges to the same ℓ (why?). We have for every $n \geq 2$

$$a_n = s_n - s_{n-1}.$$

Hence, (a_n) converges to 0. This completes the proof of Proposition 1.1.

A typical application of the divergence test is the following. Since the sequence $((-1)^n)$ does not converge to 0, the series

$$\sum_{n=1}^{\infty} (-1)^n$$

diverges.

2 Positive Terms Series

If for every $n \geq 1$, $a_n \geq 0$, then $\sum_{n=1}^{\infty} a_n$ is said to be a **positive terms series.**

Proposition 2.1 *Assume that $a_n \geq 0$ for every $n \geq 1$. Then, the series*

$$\sum_{n=1}^{\infty} a_n$$

either converges or is $+\infty$.

Proposition 2.1 implies that the symbol $\sum_{n=1}^{\infty} a_n$ is always defined for positive term series.

Proof of Proposition 2.1
For $n \geq 1$, let $s_n = \sum_{k=1}^{n} a_k$. Since $a_k \geq 0$ for every $k \geq 1$ the sequence (s_n) is increasing (why?). Hence, either (s_n) is bounded above and converges to a finite limit. Or, (s_n) is not bounded above and tends to $+\infty$. This completes the proof of Proposition 2.1.

Proposition 2.2 (Comparison Test) *Let (a_n) and (b_n) be two positive sequences with the following property. There exists a natural N such that if $n \geq N$, then $0 \leq a_n \leq b_n$.*

(i) If the series $\sum_{n=1}^{\infty} b_n$ converges so does the series $\sum_{n=1}^{\infty} a_n$.
(ii) If the series $\sum_{n=1}^{\infty} a_n = +\infty$, then $\sum_{n=1}^{\infty} b_n = +\infty$.

Proof of Proposition 2.2
For $n \geq 1$, let $s_n = \sum_{k=1}^n a_k$ and $t_n = \sum_{k=1}^n b_k$. We have for $n \geq N$

$$s_n = \sum_{k=1}^{N} a_k + \sum_{k=N+1}^{n} a_k \leq s_N + \sum_{k=N+1}^{n} b_k.$$

Note that

$$\sum_{k=N+1}^{n} b_k = t_n - t_N.$$

Therefore, for $n \geq N$

$$s_n \leq s_N - t_N + t_n. \tag{2.1}$$

Note that N is a fixed natural and therefore $s_N - t_N$ is a constant.

Assume now that the series $\sum_{n=1}^{\infty} b_n$ converges. This means that the sequence (t_n) converges and is therefore bounded. Hence, by (2.1), (s_n) is bounded above. Since (s_n) is also increasing (why?) this shows that (s_n) converges. This proves (i).

We now turn to (ii). Assume that $\sum_{n=1}^{\infty} a_n = +\infty$. Since (s_n) is increasing it cannot be bounded above (why not?). By (2.1), (t_n) is not bounded above either. Since (t_n) is increasing it must tend to $+\infty$. Hence,

$$\sum_{n=1}^{\infty} b_n = +\infty.$$

This completes the proof of (ii) and of Proposition 2.1.

Proposition 2.3 (Cauchy Condensation Test) Let (a_n) be a positive and decreasing sequence. Then, the series

$$\sum_{k=1}^{\infty} a_k$$

converges if and only if the series

$$\sum_{k=1}^{\infty} 2^k a_{2^k}$$

converges.

Proof of Proposition 2.3
For $n \geq 1$, let

$$s_n = \sum_{k=1}^{n} a_k \text{ and } t_n = \sum_{k=1}^{n} 2^k a_{2^k}.$$

For $k \geq 0$, let

$$b_k = s_{j_k} = \sum_{j=1}^{2^k} a_j.$$

Observe that (b_k) is a subsequence of (s_n). We have

$$b_k - b_{k-1} = \sum_{j=2^{k-1}+1}^{2^k} a_j.$$

Since the sequence (a_n) is decreasing we have $a_j \geq a_{2^k}$, for every natural j in $[2^{k-1} + 1, 2^k]$. Thus,

$$b_k - b_{k-1} \geq \sum_{j=2^{k-1}+1}^{2^k} a_{2^k} = (2^k - 2^{k-1})a_{2^k}.$$

Note that $2^k - 2^{k-1} = 2^{k-1}$. Therefore,

$$b_k - b_{k-1} \geq 2^{k-1} a_{2^k}.$$

We sum the preceding inequality to get

$$\sum_{k=1}^{j} (b_k - b_{k-1}) \geq \sum_{k=1}^{j} 2^{k-1} a_{2^k} = \frac{1}{2} \sum_{k=1}^{j} 2^k a_{2^k} = \frac{1}{2} t_j.$$

Since,

$$\sum_{k=1}^{j} b_k - b_{k-1} = b_j - b_0,$$

for every $j \geq 1$, we have

$$b_j - b_0 \geq \frac{1}{2} t_j.$$

Assume now that the series $\sum_{k=1}^{\infty} a_k$ converges. That is, the sequence (s_n) converges. Hence, its subsequence (b_n) converges as well. In particular, (b_n) is bounded and therefore (t_n) is bounded above. Since (t_n) is increasing, (t_n) converges. This proves the direct implication.

We now turn to the converse. Assume now that (t_n) converges. Consider again

$$b_k - b_{k-1} = \sum_{j=2^{k-1}+1}^{2^k} a_j.$$

Since the sequence (a_n) is decreasing, $a_j \le a_{2^{k-1}}$ for every natural j in $[2^{k-1}+1, 2^k]$. Thus,

$$b_k - b_{k-1} \le \sum_{j=2^{k-1}+1}^{2^k} a_{2^{k-1}} = 2^{k-1} a_{2^{k-1}}.$$

We sum the preceding inequality to get

$$\sum_{k=1}^{j}(b_k - b_{k-1}) \le \sum_{k=1}^{j} 2^{k-1} a_{2^{k-1}} = t_{j-1} - a_1.$$

Hence,

$$b_j - b_0 \le t_{j-1} - a_1.$$

Since (t_n) converges it is bounded above. Hence, (b_j) is bounded above by some m. Using that (s_n) is increasing and that $2^j > j$ for every $j \ge 1$ we have

$$s_j \le b_j \le m,$$

for every $j \ge 1$. That is, (s_n) is bounded above. Since (s_n) is increasing, it converges. This completes the proof of Proposition 2.3.

The following is an important application of the Cauchy condensation test.

Theorem 2.1 *Let $p > 0$ be a real number. The series*

$$\sum_{k=1}^{\infty} \frac{1}{k^p}$$

converges if and only if $p > 1$.

- Note that if $p \leq 1$, then

$$\sum_{k=1}^{\infty} \frac{1}{k^p} = +\infty.$$

Proof of Theorem 2.1
For $k \geq 1$, let

$$a_k = \frac{1}{k^p}.$$

For any fixed $p > 0$, the sequence (a_k) is positive and decreasing. Hence, Proposition 2.3 applies. Note that

$$a_{2^k} = \frac{1}{(2^k)^p} = (2^{-p})^k.$$

Thus,

$$2^k a_{2^k} = (2^{-p+1})^k.$$

That is, $(2^k a_{2^k})$ is a geometric sequence with ratio $r = 2^{-p+1}$. We know that the corresponding geometric series converges if and only if $|r| < 1$. Hence, the series

$$\sum_{k=1}^{\infty} \frac{1}{k^p}$$

converges if and only if $p > 1$. This completes the proof of Theorem 2.1.

Problems

1. Let $a_n \geq 0$ for all $n \geq 1$. Assume that the series $\sum_{n=1}^{\infty} a_n = +\infty$. Show that $\sum_{n=1}^{\infty} \sqrt{a_n} = +\infty$.

2.

(a) Assume that $a_n \geq 0$ for $n \geq 1$. Assume also that the series $\sum_{k=1}^{\infty} a_k$ converges. Show that $\sum_{k=1}^{\infty} a_k^2$ converges.
(b) Is the converse of (a) true?

3. Let $\sum_{k=1}^{\infty} a_k$ and $\sum_{k=1}^{\infty} b_k$ be two positive terms convergent series. Prove that $\sum_{k=1}^{\infty} a_k b_k$ converges.

4. Let $\sum_{k=1}^{\infty} a_k$ and $\sum_{k=1}^{\infty} b_k$ be two convergent series. Show that the series

$$\sum_{k=1}^{\infty}(a_k + b_k)$$

converges and that

$$\sum_{k=1}^{\infty}(a_k + b_k) = \sum_{k=1}^{\infty} a_k + \sum_{k=1}^{\infty} b_k.$$

5. Let $\sum_{k=1}^{\infty} a_k$ be a convergent series and c be a constant. Show that the series $\sum_{k=1}^{\infty} ca_k$ converges and that

$$\sum_{k=1}^{\infty} ca_k = c\sum_{k=1}^{\infty} a_k.$$

6. Let (d_n) be a sequence in $\{0, 1, 2, \ldots, 9\}$. Show that the series

$$\sum_{n=1}^{\infty} \frac{d_n}{10^n}$$

converges.

7. Let $p > 0$. Show that the series

$$\sum_{n=2}^{\infty} \frac{1}{n(\ln n)^p}$$

converges if and only if $p > 1$.

8. Consider two sequences of positive real numbers (a_n) and (b_n). Assume that

$$\lim_n \frac{a_n}{b_n} = \ell.$$

(a) Assume that $0 < \ell < +\infty$. Show that the series $\sum_{n=0}^{\infty} a_n$ and $\sum_{n=0}^{\infty} b_n$ both converge or are both infinite.
(b) Assume that $\ell = 0$. Show that if $\sum_{n=0}^{\infty} a_n = +\infty$, then $\sum_{n=0}^{\infty} b_n = +\infty$.
(c) Assume that $\ell = +\infty$. Show that if $\sum_{n=0}^{\infty} b_n = +\infty$, then $\sum_{n=0}^{\infty} a_n = +\infty$.

9. (The Number e) For $n \geq 0$ let

$$s_n = \sum_{k=0}^{n} \frac{1}{k!} = 1 + 1 + \frac{1}{2!} + \cdots + \frac{1}{n!}.$$

(a) Show that for $n \geq 2$

$$s_n < 2 + \sum_{k=1}^{n-1} \frac{1}{2^k}.$$

(b) Show that the series

$$\sum_{k=0}^{\infty} \frac{1}{k!}$$

converges.

The number e is defined as the sum of this series.

(c) Show that $2 < e < 3$.

(d) For every $n \geq 0$ show that

$$0 < e - s_n < \frac{1}{n!n}.$$

10. In this problem we show that the number e is irrational. We use the definitions and notations from Problem 9..

Assume by contradiction that $e = \frac{p}{q}$ where p and q are natural numbers.

(a) Show that $q!e$ and $q!s_q$ are natural numbers.

(b) Use Problem 9. (d) to show that

$$0 < q!(e - s_q) < \frac{1}{q}.$$

(c) Use (a) and (b) to find a contradiction.

11. In Problem 9. we defined the number e as

$$e = \sum_{n=0}^{\infty} \frac{1}{n!}.$$

In this problem we show that

$$e = \lim_n (1 + \frac{1}{n})^n.$$

For $n \geq 1$ let

$$s_n = \sum_{k=0}^{n} \frac{1}{k!} \text{ and } t_n = (1 + \frac{1}{n})^n.$$

(a) Show that for any real x we have

$$(1+x)^n = 1 + \sum_{k=1}^{n} \frac{1}{k!} n(n-1)\dots(n-k+1)x^k.$$

(b) Use (a) to show that for $n \geq 2$ we have

$$t_n = 1 + 1 + \sum_{k=2}^{n} \frac{1}{k!}(1-\frac{1}{n})(1-\frac{2}{n})\dots(1-\frac{k-1}{n}).$$

(c) Show that for all $n \geq 1$ we have $t_n \leq s_n$.
(d) Show that

$$\limsup_n t_n \leq e.$$

(e) Show that for all $2 \leq j \leq n$ we have

$$t_n \geq 1 + 1 + \sum_{k=2}^{j} \frac{1}{k!}(1-\frac{1}{n})(1-\frac{2}{n})\dots(1-\frac{k-1}{n}).$$

(f) Prove that for every $j \geq 2$,

$$\liminf_n t_n \geq \sum_{k=0}^{j} \frac{1}{k!} = s_j.$$

(g) Prove that

$$\liminf_n t_n \geq e.$$

(h) Conclude that $\lim_n t_n$ exists and that $\lim_n t_n = e$.

3 Absolute Convergence

The series $\sum_{n=1}^{\infty} a_n$ is said to **converge absolutely** if the series $\sum_{n=1}^{\infty} |a_n|$ converges. We will prove that absolute convergence implies convergence. First we state the Cauchy criterion for series.

Proposition 3.1 *Let (a_n) be a sequence of real numbers. Then, the series $\sum_{n=1}^{\infty} a_n$ converges if and only for every $\epsilon > 0$ there exists a natural N such that if $n > p \geq N$, then*

$$\left| \sum_{k=p+1}^{n} a_k \right| < \epsilon.$$

Proof of Proposition 3.1
For $n \geq 1$ let $s_n = \sum_{k=1}^{n} a_k$. The series $\sum_{k=1}^{\infty} a_k$ converges if and only if the sequence (s_n) converges. This sequence converges if and only it is Cauchy (see Theorem 6.1 in Chapter 2). The sequence (s_n) is Cauchy if and only if for every $\epsilon > 0$ there is a natural N such that if $n > p \geq N$, then

$$|s_n - s_p| = \left| \sum_{k=p+1}^{n} a_k \right| < \epsilon.$$

This completes the proof of Proposition 3.1.

Theorem 3.1 *If a series converges absolutely, then it converges.*

The converse of Theorem 3.1 is not true. We will show that $\sum_{n\geq 1} \frac{(-1)^n}{n}$ converges but not absolutely (why not?).

Proof of Theorem 3.1
Let (a_n) be a sequence of real numbers. For $n \geq 1$ let

$$s_n = \sum_{k=1}^{n} a_k \text{ and } t_n = \sum_{k=1}^{n} |a_k|.$$

We want to show that if (t_n) converges, so does (s_n). By the Cauchy criterion we have that if (t_n) converges, then for every $\epsilon > 0$ there is a natural N such that if $n > p \geq N$,

$$|t_n - t_p| < \epsilon.$$

Note that

$$|t_n - t_p| = \left| \sum_{k=p+1}^{n} |a_k| \right| = \sum_{k=p+1}^{n} |a_k| < \epsilon.$$

On the other hand, by the triangle inequality, for $n > p \geq N$

$$|s_n - s_p| = \left| \sum_{k=p+1}^{n} a_k \right| \leq \sum_{k=p+1}^{n} |a_k| < \epsilon.$$

This proves that (s_n) is a Cauchy sequence. Therefore, (s_n) converges. This completes the proof of Theorem 3.1.

We now turn to two important tests for absolute convergence.

Proposition 3.2 (Root Test) *Let (a_n) be a sequence of real numbers.*

(i) *If $\limsup_n |a_n|^{1/n} < 1$, then the series $\sum_{n\geq 1} a_n$ converges absolutely.*
(ii) *If $\limsup_n |a_n|^{1/n} > 1$, then the series $\sum_{n\geq 1} a_n$ does not converge.*

- Note that if $\limsup_n |a_n|^{1/n} = 1$, the root test does not give any information on the convergence of the series. To see this consider the following example. Since the sequence $(n^{1/n})$ converges to 1 we have that

$$(\frac{1}{n^p})^{1/n}$$

also converges to 1 for any real p. However, the series $\sum_{n\geq 1} \frac{1}{n^p}$ is infinite for $p = 1$ and converges for $p = 2$.
- When we say that the series $\sum_{n\geq 1} a_n$ does not converge. It means that the limit of partial sums is not finite. The limit may be infinite or not exist at all.

Proof of Proposition 3.2
Let $\limsup_n |a_n|^{1/n} = \bar{\ell}$.

We first prove (i). Assume that $\bar{\ell} < 1$. Let $\bar{\ell} < s < 1$. There is a natural N such that if $n \geq N$, then

$$|a_n|^{1/n} < s,$$

see Proposition 2.6 in Chapter 3. Hence, for every $n \geq N$, $|a_n| < s^n$. Since the series $\sum_{n\geq 1} s^n$ converges (why?), by the Comparison test the series $\sum_{n\geq 1} |a_n|$ converges as well. This proves (i).

We now turn to (ii). Since $\limsup_n(|a_n|^{1/n}) = \bar{\ell}$ there is a subsequence $(|a_{n_k}|^{1/n_k})$ that tends to $\bar{\ell}$ (which may be finite or infinite), see Proposition 2.3 in Chapter 3.

If $\bar{\ell}$ is finite, let

$$\epsilon = \frac{\bar{\ell} - 1}{2} > 0.$$

There exists a natural N such that if $k \geq N$, then

$$\left| |a_{n_k}|^{1/n_k} - \bar{\ell} \right| < \epsilon.$$

Therefore

$$|a_{n_k}|^{1/n_k} > \frac{\bar{\ell} + 1}{2} > 1.$$

This shows that the subsequence $(|a_{n_k}|)$ tends to $+\infty$ (why?).

If $\bar{\ell} = +\infty$, let $b > 1$, then there exists a natural N_1 such that if $k \geq N$

$$|a_{n_k}|^{1/n_k} > b.$$

Hence, $(|a_{n_k}|)$ tends to $+\infty$.

In both cases, (a_n) does not converge to 0. By the divergence test the series $\sum_{n\geq 1} a_n$ does not converge. This completes the proof of Proposition 3.2.

Proposition 3.3 (Ratio Test) *Let (a_n) be a sequence of nonzero real numbers.*

(i) *If*

$$\limsup_n \frac{|a_{n+1}|}{|a_n|} < 1,$$

then the series $\sum_{n\geq 1} a_n$ converges absolutely.

(ii) *If*

$$\liminf_n \frac{|a_{n+1}|}{|a_n|} > 1,$$

then the series $\sum_{n\geq 1} a_n$ does not converge.

- It turns out (see the problems) that the root test is strictly better than the ratio test in the following sense. Every time the root test is inconclusive, so is the ratio test. But there are examples for which the ratio test is inconclusive but the root test is conclusive. However, many times the limit of a ratio is easier to compute than the limit of a root. This is why both tests are useful.

Proof of Proposition 3.3

We first prove (i). Let

$$\limsup_n \frac{|a_{n+1}|}{|a_n|} = \bar{\ell} < 1.$$

Let $\bar{\ell} < b < 1$. There is a natural N such that if $n \geq N$, then

$$\frac{|a_{n+1}|}{|a_n|} < b,$$

see Proposition 2.6 in Chapter 3. Hence, for $n \geq N$

$$|a_{n+1}| < b|a_n|.$$

An easy induction shows that for any $k \geq 1$,

$$|a_{N+k}| < b^k |a_N|.$$

By letting $n = N + k$ we may rewrite this inequality as

$$|a_n| < b^{n-N}|a_N| = \frac{|a_N|}{b^N}b^n,$$

for all $n \geq N$. The series with general term $\frac{|a_N|}{b^N}b^n$ converges (why?). Hence, by the comparison test the series $\sum_{n\geq 1}|a_n|$ converges as well. This proves (i).

We now turn to (ii). Let

$$\underline{\ell} = \liminf_n \frac{|a_{n+1}|}{|a_n|} > 1.$$

Let $1 < c < \underline{\ell}$. By Proposition 2.6 in Chapter 3 there exists an N such that if $n \geq N$, then

$$\frac{|a_{n+1}|}{|a_n|} > c.$$

That is, for all $n \geq N$

$$|a_{n+1}| > c|a_n|.$$

As above an easy induction shows for every $k \geq 1$

$$|a_{N+k}| > c^k|a_N|.$$

Therefore, for all $n \geq N$

$$|a_n| > c^{n-N}|a_N|.$$

Since $c > 1$ and $|a_N| > 0$ this shows that $(|a_n|)$ tends to infinity. In particular, (a_n) does not converge to 0. By the divergence test the series $\sum_{n\geq 1} a_n$ diverges. This completes the proof of Proposition 3.3.

4 Conditional Convergence

A series $\sum_{n\geq 1} a_n$ is said to be **conditionally convergent** if $\sum_{n\geq 1} a_n$ converges but

$$\sum_{n\geq 1}|a_n| = +\infty.$$

The following test will be useful in providing examples of conditional convergence.

Theorem 4.1 (Abel's Convergence Test) *Let (b_n) be a positive decreasing sequence that converges to 0. Let (a_n) be a sequence and for $n \geq 1$ let*

$$s_n = \sum_{k=1}^{n} a_k.$$

Assume that the sequence (s_n) is bounded. Then, the series $\sum_{n\geq 1} a_n b_n$ converges.

- Note that multiple hypotheses must be checked before applying Abel's test. Therefore it is not a widely applicable test. It is however useful to show convergence for series of the type $\sum_{n\geq 1}(-1)^n b_n$ and $\sum_{n\geq 1} \frac{\sin n}{n}$ as we will see below.

Proof of Theorem 4.1
Note that $a_n = s_n - s_{n-1}$ for every $n \geq 1$. Hence, for $n > m$ we have

$$\sum_{j=m+1}^{n} a_j b_j = \sum_{j=m+1}^{n} (s_j - s_{j-1}) b_j = \sum_{j=m+1}^{n} s_j b_j - \sum_{j=m+1}^{n} s_{j-1} b_j.$$

Shifting the index in the second sum we get

$$\sum_{j=m+1}^{n} a_j b_j = \sum_{j=m+1}^{n} s_j b_j - \sum_{j=m}^{n-1} s_j b_{j+1} = \sum_{j=m+1}^{n-1} s_j (b_j - b_{j+1}) + s_n b_n - s_m b_{m+1}.$$

Hence,

$$\begin{aligned} \left| \sum_{j=m+1}^{n} a_j b_j \right| &= \left| \sum_{j=m+1}^{n-1} s_j (b_j - b_{j+1}) + s_n b_n - s_m b_{m+1} \right| \\ &\leq \sum_{j=m+1}^{n-1} |s_j||b_j - b_{j+1}| + |s_n||b_n| + |s_m||b_{m+1}|. \end{aligned}$$

By assumption, the sequence (s_n) is bounded by some k. Therefore,

$$\left| \sum_{j=m+1}^{n} a_j b_j \right| \leq k \sum_{j=m+1}^{n-1} |b_j - b_{j+1}| + k|b_n| + k|b_{m+1}|.$$

We now use the properties of the sequence (b_n). Since (b_n) is decreasing,

$$|b_j - b_{j+1}| = b_j - b_{j+1}$$

for every $j \geq 1$. Hence,

$$\sum_{j=m+1}^{n-1} |b_j - b_{j+1}| = \sum_{j=m+1}^{n-1} (b_j - b_{j+1}) = b_{m+1} - b_n.$$

Using also that $b_j \geq 0$ for all $j \geq 1$ we get,

$$\left| \sum_{j=m+1}^{n} a_j b_j \right| \leq k(b_{m+1} - b_n + b_n + b_{m+1}) = 2kb_{m+1}.$$

Since (b_n) converges to 0 there exists an N such that if $j \geq N$, then

$$b_j = |b_j| < \frac{\epsilon}{2k}.$$

Hence, for $n > m \geq N$ we have

$$\left| \sum_{j=m+1}^{n} a_j b_j \right| < 2k \frac{\epsilon}{2k} = \epsilon.$$

By Proposition 3.1 the series $\sum_{n\geq 1} a_n b_n$ converges. This completes the proof of Theorem 4.1.

The following is an application of Theorem 4.1.

Theorem 4.2 (Alternating Series Test) *Let (b_n) be a positive decreasing sequence that converges to 0. Then, the series $\sum_{n\geq 1}(-1)^n b_n$ converges.*

Proof of Theorem 4.2
For $n \geq 1$ let

$$s_n = \sum_{k=1}^{n} (-1)^k.$$

To apply Theorem 4.1 we only need to check that the sequence (s_n) is bounded.
Summing the geometric sequence we get for all $n \geq 1$

$$s_n = -\frac{(-1)^n - 1}{-1 - 1} = \frac{1}{2}((-1)^n - 1).$$

This shows that $s_n = -1$ for n odd and $s_n = 0$ for n even. In particular, (s_n) is bounded. By Theorem 4.1 the series $\sum_{n\geq 1}(-1)^n b_n$ converges. This completes the proof of Theorem 4.2.

Example 4.1 Let $p > 0$. Consider the series

$$\sum_{n\geq 1} \frac{(-1)^n}{n^p}.$$

For which p is this series conditionally convergent?

The series is absolutely convergent if and only if $p > 1$. Consider $0 < p \leq 1$. For $n \geq 1$ let $b_n = \frac{1}{n^p}$. Clearly, (b_n) is decreasing and converges to 0. Theorem 4.2 applies. The series $\sum_{n\geq 1} \frac{(-1)^n}{n^p}$ converges. Thus, this series is conditionally convergent if and only if $0 < p \leq 1$.

5 Rearrangements

Let σ be a function from $\mathbb{N}$ to $\mathbb{N}$ with the following property. The sequence $(\sigma(n))$ is a sequence of natural numbers for which every natural number appears once and only once. Then, the series

$$\sum_{n\geq 1} a_{\sigma(n)}$$

is called a **rearrangement** of the series $\sum_{n\geq 1} a_n$. Note that a rearranged series has exactly the same terms as the original series. The only difference is the order in which the two series are summed. If the sums were finite the summation order would not matter (addition is a commutative operation!). However, with infinite sums a rearranged series may be quite different from the original series. Next we give an example.

Example 5.1 Consider the **harmonic** series

$$\sum_{k=1}^{\infty} \frac{(-1)^{k+1}}{k} = 1 - \frac{1}{2} + \frac{1}{3} - \frac{1}{4} \dots .$$

It is easy to check that Theorem 4.2 applies (just note that $(-1)^{k+1} = -(-1)^k$) and that this series converges. In fact the sum of the harmonic series is $\ln 2$, see for instance Section 4.1 in Schinazi (2011). Here we only need to know that the sum is not 0.

Consider now the following rearrangement of the harmonic series.

$$1 - \frac{1}{2} - \frac{1}{4} + \frac{1}{3} - \frac{1}{6} - \frac{1}{8} + \frac{1}{5} - \frac{1}{10} - \frac{1}{12} \dots = \sum_{n=0}^{\infty} (\frac{1}{2n+1} - \frac{1}{4n+2} - \frac{1}{4n+4}).$$

This is indeed a rearrangement of the harmonic series: every term of the harmonic series appears once and only once in the rearranged series. The general term of the rearranged series is for $n \geq 0$

$$\left(\frac{1}{2n+1} - \frac{1}{4n+2}\right) - \frac{1}{4n+4} = \frac{1}{4n+2} - \frac{1}{4n+4} = \frac{1}{2}\left(\frac{1}{2n+1} - \frac{1}{2n+2}\right).$$

Hence, the rearranged series is

$$\frac{1}{2}\sum_{n=0}^{\infty}\left(\frac{1}{2n+1} - \frac{1}{2n+2}\right) = \frac{1}{2}\sum_{n=1}^{\infty}\frac{(-1)^{n+1}}{n}.$$

That is, the rearranged sum is one half of the original sum! Since the sum is nonzero we see that the rearrangement is changing the infinite sum.

Example 5.1 is an example of a much more general phenomenon as we see next.

Theorem 5.1 *Assume that the series $\sum_{n\geq 1} a_n$ is conditionally convergent. For any $-\infty \leq \ell \leq +\infty$ there exists a rearrangement of $\sum_{n\geq 1} a_n$ which is equal to ℓ.*

We will not prove Theorem 5.1. See the problems or Rudin (1976).

The next result shows that things are a lot more stable when the series converges absolutely.

Theorem 5.2 *Assume that the series $\sum_{n\geq 1} a_n$ converges absolutely. For any rearrangement σ, we have*

$$\sum_{n\geq 1} a_{\sigma(n)} = \sum_{n\geq 1} a_n.$$

In words, if a series converges absolutely rearrangements do not affect its infinite sum.

Proof of Theorem 5.2

The series $\sum_{k=1}^{\infty} |a_k|$ converges, so by Proposition 3.1 there is a natural N such that for $n > m \geq N$

$$\sum_{k=m+1}^{n} |a_k| < \epsilon.$$

By letting n go to infinity we get

$$\sum_{k=m+1}^{\infty} |a_k| \leq \epsilon. \tag{5.1}$$

The function σ is one-to-one. Thus, if p is large enough, then

$$\{1, 2, \ldots, N\} \subset \{\sigma(1), \sigma(2), \ldots, \sigma(p)\}.$$

Observe that $p \geq N$. For $n \geq 1$, let $s_n = \sum_{k=1}^{n} a_k$ and $s'_n = \sum_{k=1}^{n} a_{\sigma(k)}$. Note now that if $n \geq p$, then the sum s'_n has all the terms $a_1, a_2, \ldots, a_N$ and so does the sum s_n. Hence,

$$|s_n - s'_n| \leq \sum_{k=N+1}^{\infty} |a_k|.$$

By (5.1) we get that if $n \geq p$, then $|s_n - s'_n| \leq \epsilon$. This shows that (s'_n) converges and has the same limit as (s_n). The proof of Theorem 5.2 is complete.

Corollary 5.1 *Let (a_n) be a positive sequence. For any rearrangement σ we have*

$$\sum_{n\geq 1} a_{\sigma(n)} = \sum_{n\geq 1} a_n.$$

Proof of Corollary 5.1
There are two cases.

- If $\sum_{n\geq 1} a_n < +\infty$, then the series converges absolutely. Theorem 5.2 applies and we are done.
- If $\sum_{n\geq 1} a_n = +\infty$, assume by contradiction that there exists a σ such that

$$\sum_{n\geq 1} a_{\sigma(n)} < +\infty.$$

Hence, the rearranged series converges absolutely. Theorem 5.2 applies. We can think of the original series $\sum_{n\geq 1} a_n$ as a rearrangement of $\sum_{n\geq 1} a_{\sigma(n)}$ (why?). Thus, by Theorem 5.2 we have

$$\sum_{n\geq 1} a_n = \sum_{n\geq 1} a_{\sigma(n)} < +\infty.$$

We have a contradiction. Hence, $\sum_{n\geq 1} a_{\sigma(n)} = +\infty$. This completes the proof of Corollary 5.1.

Problems

1. Assume that $\lim_n |a_n|^{1/n} = \ell$ exists.

(a) Show that if $\ell < 1$, then the series $\sum_{n\geq 1} a_n$ converges absolutely.
(b) Show that if $\ell > 1$, then the series $\sum_{n\geq 1} a_n$ does not converge.

2. Assume that

$$\lim_n \frac{|a_{n+1}|}{|a_n|} = \ell$$

exists.

(a) Show that if $\ell < 1$, then the series $\sum_{n\geq 1} a_n$ converges absolutely.
(b) Show that if $\ell > 1$, then the series $\sum_{n\geq 1} a_n$ does not converge.

3. Let (a_n) be defined as follows. For every $k \geq 1$ let

$$a_{2k-1} = \frac{1}{3^k} \text{ and } a_{2k} = \frac{1}{2^k}.$$

(a) Show that

$$\limsup_n a_n^{1/n} = \frac{1}{\sqrt{2}}.$$

(b) Show that

$$\limsup_n \frac{a_{n+1}}{a_n} = +\infty.$$

(c) Show that

$$\liminf_n \frac{a_{n+1}}{a_n} = 0.$$

(d) Apply the root test and the ratio test.

4. Let (a_n) be a sequence of strictly positive numbers. Let

$$\bar{\ell} = \limsup_n \frac{a_{n+1}}{a_n}.$$

Assume that $\bar{\ell}$ is finite.

(a) Show that for any $b > \bar{\ell}$ there exists a natural N such that $a_{n+1} < ba_n$ for $n \geq N$.
(b) Show that for every $n \geq N$ we have

$$a_n \leq a_N b^{n-N}.$$

(c) Show that

$$\limsup_n a_n^{1/n} \leq b.$$

(d) Prove that $\limsup_n a_n^{1/n} \leq \bar{\ell}$.
(e) Explain why (d) holds for $\bar{\ell} = +\infty$ as well.

5. Let (a_n) be a sequence of strictly positive numbers. Let

$$\underline{\ell} = \liminf_n \frac{a_{n+1}}{a_n}.$$

Assume that $\underline{\ell}$ is finite. Show that

$$\liminf_n a_n^{1/n} \geq \underline{\ell}.$$

(Follow the pattern of Problem 4..)

6. Using Problems 4. and 5. show that if the ratio test is conclusive so is the root test.

7.

(a) Let j be a natural number. Show that

$$\sin j = \frac{1}{2\sin(1/2)}\left(\cos(j - 1/2) - \cos(j + 1/2)\right).$$

(b) For $n \geq 1$, let $s_n = \sum_{j=1}^{n} \sin j$. Show that

$$s_n = \frac{1}{2\sin(1/2)}\left(\cos(1/2) - \cos(n + 1/2)\right).$$

(c) Show that the series

$$\sum_{j \geq 1} \frac{\sin j}{j}$$

converges.

8. Let (b_n) be a decreasing sequence converging to 0. By Theorem 4.2 we know that the series $\sum_{n\geq 1}(-1)^n b_n$ converges. Let ℓ be this infinite sum. For $n \geq 1$, let

$$d_n = \sum_{j=1}^{n}(-1)^j b_j.$$

(a) Show that the subsequence (d_{2n}) is decreasing.
(b) Show that the subsequence (d_{2n+1}) is increasing.
(c) Show that for every $n \geq 1$ we have

$$d_{2n+1} \leq \ell \leq d_{2n}.$$

(d) Apply (c) to give a lower bound and an upper bound for

$$\sum_{n\geq 1} \frac{(-1)^{n+1}}{n}.$$

9. Let (a_n) be a sequence of real numbers. For every $n \geq 1$ define

$$a_n^+ = \max(a_n, 0) \text{ and } a_n^- = \max(-a_n, 0).$$

(a) Show that (a_n^+) and (a_n^-) are positive terms sequences.
(b) Show that for every $n \geq 1$

$$a_n = a_n^+ - a_n^-.$$

(c) Show that for every $n \geq 1$

$$|a_n| = a_n^+ + a_n^-.$$

10. We use the notation and results of Problem 9..

(a) Show that the series $\sum_{n\geq 1} a_n$ is absolutely convergent if and only if

$$\sum_{n\geq 1} a_n^+ < +\infty \text{ and } \sum_{n\geq 1} a_n^- < +\infty.$$

(b) Show that if the series $\sum_{n\geq 1} a_n$ is conditionally convergent, then

$$\sum_{n\geq 1} a_n^+ = +\infty \text{ and } \sum_{n\geq 1} a_n^- = +\infty.$$

(c) For any fixed r show that a conditionally convergent series may be rearranged in such a way its partial sums are alternatively smaller and larger than r. (This is the first step in the proof of Theorem 5.1.Use (b)).

Chapter 5
Convergence of Functions

Let (f_n) be a sequence of functions converging to some function f. Here are a few questions we are interested in.

If the f_n are continuous, is f continuous?

If the f_n are differentiable, is f differentiable?

Is it true that

$$\lim_{n\to\infty}\int_a^b f_n dx = \int_a^b \lim_{n\to\infty} f_n dx?$$

We will give some partial answers to these questions in this chapter. First we need to define what we mean by convergence of a sequence of functions. There are many different ways a sequence of functions can converge. In this chapter we will concentrate on two of them.

1 Pointwise and Uniform Convergence

Consider a sequence (f_n) of real valued functions all defined on the same set $S \subset \mathbb{R}$. The sequence (f_n) is said to **converge pointwise** to f on S if for every x in S,

$$\lim_{n\to\infty} f_n(x) = f(x).$$

For every fixed x, $(f_n(x))$ is a numerical sequence. Hence, (f_n) converges pointwise to f on S means that for every x in S and every $\epsilon > 0$ there exists N (that depends on ϵ and x) such that if $n \geq N$, then $|f_n(x) - f(x)| < \epsilon$.

Example 1.1 For $n \geq 1$, let $f_n(x) = e^{-nx}$ and $S = [0, \infty)$. Does the sequence (f_n) converge?

R. B. Schinazi, *From Classical to Modern Analysis*,
https://doi.org/10.1007/978-3-319-94583-5_5

Let $x \geq 0$. We have

$$f_n(x) = e^{-nx} = (e^{-x})^n = r^n$$

where $r = e^{-x}$. There are two cases. If $x = 0$, then $r = 1$ and $r^n = 1$ for all $n \geq 1$. Therefore $(f_n(0))$ converges to 1. On the other hand, if $x > 0$, then $r < 1$ and (r^n) converges to 0. That is, for $x > 0$, $(f_n(x))$ converges to 0. In summary, we have proved that the sequence of functions (f_n) converges pointwise to a function f on $[0, \infty)$ where f is defined by $f(0) = 1$ and $f(x) = 0$ for $x > 0$.

- Example 1.1 shows that a sequence (f_n) of continuous functions may converge pointwise to a limit f which is not continuous!

Example 1.2 Consider the sequence (f_n) defined on [0, 1] by

$$f_n(0) = 0$$

$$f_n(x) = n \text{ for } x \in (0, \frac{1}{n}]$$

$$f_n(x) = 0 \text{ for } x \in (\frac{1}{n}, 1].$$

We first show that (f_n) converges pointwise to 0 on [0, 1]. If $x = 0$, then $(f_n(0))$ is identically 0 and therefore converges to 0. Let x be a fixed number in (0, 1]. There exists N such that $N > \frac{1}{x}$ (why?) and therefore $x > \frac{1}{N}$. Hence, for all $n \geq N$, $x > \frac{1}{n}$. Therefore, $f_n(x) = 0$ for all $n \geq N$. Thus, $(f_n(x))$ converges to 0. We have proved that (f_n) converges pointwise to the 0 function on [0, 1].

Example 1.3 Consider the sequence (f_n) introduced in Example 1.2. Is it true that

$$\lim_{n\to\infty} \int_0^1 f_n dx = \int_0^1 \lim_{n\to\infty} f_n dx?$$

First note that each f_n is Riemann integrable. This is so because for every $n \geq 1$, f_n is bounded on [0, 1] and continuous except at $x = 0$ and $x = 1/n$. We have

$$\int_0^1 f_n(x)dx = \int_0^{1/n} ndx = 1.$$

Hence, the sequence of integrals $(\int_0^1 f_n(x)dx)$ is identically 1 and therefore converges to 1. On the other hand, we have shown in Example 1.2 that (f_n) converges pointwise to the null function. Hence,

$$\int_0^1 \lim_{n\to\infty} f_n dx = \int_0^1 0dx = 0.$$

This is an example for which we cannot interchange the limit and the integral.

- The preceding examples show that pointwise convergence is not enough for our purposes. A sequence of continuous functions need not converge to a continuous function and the interchange of limit and integral need not be true. This is why we now turn to a more stringent type of convergence.

Consider a sequence of functions (f_n) all defined on the same set $S \subset \mathbb{R}$. The sequence (f_n) is said to **converge uniformly** to f on S if for every $\epsilon > 0$ there exists a natural N such that for all $n \geq N$ we have

$$|f_n(x) - f(x)| < \epsilon \text{ for all } x \in S.$$

The critical difference between pointwise and uniform convergence is the following. In the case of uniform convergence it is the **same** N for all x in S. This is why it is called **uniform** convergence. In the case of pointwise convergence for each x in S we have a possibly different N. The following criterion will be useful in checking for uniform convergence.

Proposition 1.1 (Uniform Convergence Criterion) *Consider the sequence of functions (f_n) defined on S. For a fixed function f defined on S and for $n \geq 1$ let*

$$m_n = \sup\{|f_n(x) - f(x)| : x \in S\}.$$

The sequence of functions (f_n) converges uniformly to f if and only if the numerical sequence (m_n) converges to 0.

As always, sup stands for least upper bound. With no further assumptions on (f_n) and f there is no guarantee that the set $\{|f_n(x) - f(x)| : x \in S\}$ is bounded above. If for some n the set is not bounded above we set $m_n = +\infty$.

Proof of Proposition 1.1
Assume first that (f_n) converges uniformly to f on S. Let $\epsilon > 0$, there exists a natural N such that for all $n \geq N$ we have

$$|f_n(x) - f(x)| < \epsilon \text{ for all } x \in S.$$

This shows that for $n \geq N$ the set

$$A_n = \{|f_n(x) - f(x)| : x \in S\}$$

is bounded above by ϵ. Hence the supremum (i.e., the least upper bound) m_n of this set exists (why?) and is less than the upper bound ϵ. That is, for $n \geq N$ we have $m_n = |m_n - 0| \leq \epsilon$. This proves that (m_n) converges to 0. The direct implication is proved.

We now prove the converse. Assume that (m_n) converges to 0. Let $\epsilon > 0$, there is a natural N such that if $n \geq N$, then $|m_n - 0| < \epsilon$. Since m_n is positive we have $m_n < \epsilon$. That is, the least upper bound m_n of the set A_n is less than ϵ for $n \geq N$. Hence, ϵ is an upper bound for A_n. Therefore, for $n \geq N$ we have

$$|f_n(x) - f(x)| \leq \epsilon$$

for all x in S. This proves that (f_n) converges uniformly to f on S. The proof of Proposition 1.1 is complete.

It is easy to see that uniform convergence implies pointwise convergence. We will see shortly that the converse is not true. The next result will be useful in finding a uniform limit if any.

Proposition 1.2 *Assume that (f_n) converges uniformly to f and pointwise to g on a set S. Then, $f = g$ on S.*

Proof of Proposition 1.2
Note that for any fixed x_0 in S we have

$$|f_n(x_0) - f(x_0)| \leq \sup_{x \in S} |f_n(x) - f(x)| = m_n.$$

Since (f_n) converges uniformly to f, (m_n) converges to 0. Therefore, $(f_n(x_0))$ converges to $f(x_0)$.

Using now that (f_n) converges pointwise to g, $(f_n(x_0))$ also converges to $g(x_0)$. Since limits are unique we must have $f(x_0) = g(x_0)$. This shows that $f = g$ on S and completes the proof of Proposition 1.2.

Proposition 1.2 suggests a strategy to find a uniform limit (if there is one). We first look for a pointwise limit. If there is one, then we use the uniform criterion to check whether the convergence is uniform. Next we use this strategy.

Example 1.4 For $n \geq 1$, let f_n be defined by

$$f_n(x) = \sqrt{x^2 + \frac{1}{n}}.$$

Does (f_n) converge uniformly on $\mathbb{R}$?

We first compute the pointwise limit (if any). Let x be a fixed real number. As n goes to infinity the numerical sequence $(x^2 + \frac{1}{n})$ converges to x^2. Since the square root function is continuous on $[0, \infty)$ we get that $(f_n(x))$ converges to $\sqrt{x^2} = |x|$. That is the sequence (f_n) converges pointwise on $\mathbb{R}$ to the function f defined by $f(x) = |x|$.

Now that we have a function f the second step is to estimate m_n. We compute

$$|f_n(x) - f(x)| = |\sqrt{x^2 + \frac{1}{n}} - \sqrt{x^2}| = \frac{\frac{1}{n}}{\sqrt{x^2 + \frac{1}{n}} + \sqrt{x^2}}.$$

Since $x^2 \geq 0$ we have for all x in $\mathbb{R}$

$$\sqrt{x^2 + \frac{1}{n}} + \sqrt{x^2} \geq \sqrt{\frac{1}{n}}.$$

Note that the preceding inequality is an equality at $x = 0$. For all x in $\mathbb{R}$,

$$|f_n(x) - f(x)| \leq \frac{\frac{1}{n}}{\frac{1}{\sqrt{n}}} = \frac{1}{\sqrt{n}}.$$

The value $\frac{1}{\sqrt{n}}$ is attained at $x = 0$. This shows that $\frac{1}{\sqrt{n}}$ is the maximum (and therefore the least upper bound) of the set $\{|f_n(x) - f(x)| : x \in \mathbb{R}\}$. Hence, $m_n = \frac{1}{\sqrt{n}}$ for $n \geq 1$. The sequence (m_n) converges to 0 and by the uniform convergence criterion (f_n) converges uniformly to f on $\mathbb{R}$.

2 Uniform Convergence and Continuity

We now show that uniform convergence preserves continuity.

Theorem 2.1 *Assume that the sequence (f_n) converges uniformly to f on S. Assume that for all $n \geq 1$ the function f_n is continuous at $a \in S$. Then f is also continuous at a.*

Proof of Theorem 2.1
By the uniform convergence criterion the sequence (m_n) converges to 0. Hence, for $\epsilon > 0$ there is a natural N such that if $n \geq N$, then

$$|m_n - 0| = m_n < \epsilon/3.$$

Let p be a fixed natural number larger than or equal to N. We have

$$|f(x) - f(a)| = |f(x) - f_p(x) + f_p(x) - f_p(a) + f_p(a) - f(a)|.$$

By the triangle inequality

$$|f(x) - f(a)| \leq |f(x) - f_p(x)| + |f_p(x) - f_p(a)| + |f_p(a) - f(a)|.$$

Recall now that m_n is the least upper bound of the set $\{|f_n(x) - f(x)| : x \in S\}$, therefore $|f(x) - f_p(x)|$ and $|f_p(a) - f(a)|$ are both less than m_p. Hence,

$$|f(x) - f(a)| \leq m_p + |f_p(x) - f_p(a)| + m_p.$$

Since $p \geq N$ we know that $m_p < \epsilon/3$. Using that f_p is continuous at a there exists a $\delta > 0$ such that if $x \in S$ and $|x - a| < \delta$, then

$$|f_p(x) - f_p(a)| < \epsilon/3.$$

That is,

$$|f(x) - f(a)| \leq \epsilon/3 + \epsilon/3 + \epsilon/3 = \epsilon$$

for x in S such that $|x - a| < \delta$. This proves that f is continuous at a and completes the proof of Theorem 2.1.

Example 2.1 We revisit example 1.1. Let $f_n(x) = e^{-nx}$ for $n \geq 1$ and x in $S = [0, \infty)$. We showed that (f_n) converges to f pointwise on S where f is defined by $f(0) = 1$ and $f(x) = 0$ for $x > 0$. Is the convergence uniform?

Each f_n is clearly continuous on S, however f is not. Hence, the convergence cannot be uniform. If the convergence was uniform the limit function f would be continuous according to Theorem 2.1. Therefore, the sequence (f_n) converges pointwise to f but not uniformly. This example shows that pointwise convergence does not imply uniform convergence.

We now state a Cauchy criterion for uniform convergence. We will need it in several applications. A sequence (f_n) of functions is said to be **uniformly Cauchy** on S if for every $\epsilon > 0$ there exists a natural N such that if $n > p \geq N$, then for every x in S

$$|f_n(x) - f_p(x)| < \epsilon.$$

Theorem 2.2 *A sequence (f_n) converges uniformly on S if and only if it is uniformly Cauchy on S.*

Proof of Theorem 2.2

Assume that the sequence (f_n) converges uniformly to some f on S. Then, for every $\epsilon > 0$ there exists a natural number N such that if $n \geq N$, then for every x in S,

$$|f_n(x) - f(x)| < \frac{\epsilon}{2}.$$

Let $n > p \geq N$, we have for every x in S,

$$|f_n(x) - f_p(x)| = |f_n(x) - f(x) + f(x) - f_p(x)|$$

$$\leq |f_n(x) - f(x)| + |f(x) - f_p(x)|$$
$$< \frac{\epsilon}{2} + \frac{\epsilon}{2} = \epsilon.$$

That is, the sequence (f_n) is uniformly Cauchy. This proves the direct implication.

We now turn to the converse. Assume that the sequence (f_n) is uniformly Cauchy. There exists N such that if $n > p \geq N$, then for every x in S

$$|f_n(x) - f_p(x)| < \epsilon.$$

This shows that for any fixed x the sequence $(f_n(x))$ is a Cauchy sequence in the real numbers. But a Cauchy sequence in $\mathbb{R}$ converges (see Theorem 6.1 in Chapter 2). Therefore, for every x in S, $(f_n(x))$ converges to some $f(x)$. This defines a function f on S.

It remains to prove that (f_n) converges to f uniformly. We use again that for $n > p \geq N$

$$|f_n(x) - f_p(x)| < \epsilon,$$

for every x in S. We let n go to infinity in the inequality to get

$$|f(x) - f_p(x)| \leq \epsilon,$$

for $p \geq N$ and every x in S. The critical observation is that it is the same N for every x. This proves that (f_n) converges uniformly to f on S and completes the proof of Theorem 2.2.

Putting together Theorems 2.1 and 2.2 we get the following important corollary.

Corollary 2.1 *Assume that for all $n \geq 1$, f_n is continuous on S. Assume that (f_n) is uniformly Cauchy on S. Then, (f_n) converges uniformly to a function f. Moreover, f is continuous on S.*

3 Uniform Convergence and Integration

We assume that the reader is familiar with the Riemann integral. All the facts we use are elementary and can be found in a calculus book. See for instance Schinazi (2011).

Theorem 3.1 *Assume that for each $n \geq 1$ the function f_n is continuous on the closed bounded interval $[a, b]$. Assume also that the sequence (f_n) converges uniformly to f on $[a, b]$. Then*

$$\lim_{n\to\infty}\int_a^b f_n(x)dx = \int_a^b \lim_{n\to\infty} f_n(x) = \int_a^b f(x)dx.$$

In words, if the sequence of continuous functions converges uniformly, then we may interchange limit and integral.

Proof of Theorem 3.1
Given that the functions f_n are assumed to be continuous, they are Riemann integrable. Since uniform convergence preserves continuity f is continuous and therefore integrable as well.

Let $\epsilon > 0$, by the uniform convergence criterion there is a natural N such that if $n \geq N$, then $|m_n - 0| = m_n < \epsilon/(b-a)$. Now consider

$$|\int_a^b f_n(x)dx - \int_a^b f(x)dx| = |\int_a^b (f_n(x) - f(x))dx| \leq \int_a^b |f_n(x) - f(x)|dx$$

where the last inequality is the well-known inequality

$$|\int_a^b g(x)dx| \leq \int_a^b |g(x)|dx$$

valid for any Riemann integrable function g. By definition of m_n we have for every x in $[a, b]$

$$|f_n(x) - f(x)| \leq m_n$$

and therefore

$$\int_a^b |f_n(x) - f(x)|dx \leq \int_a^b m_n dx = m_n(b-a).$$

For $n \geq N$ we have $m_n < \epsilon/(b-a)$. Hence,

$$|\int_a^b f_n(x)dx - \int_a^b f(x)dx| \leq \int_a^b |f_n(x) - f(x)|dx \leq m_n(b-a) < \epsilon.$$

This proves that $\lim_{n\to\infty}\int_a^b f_n(x)dx = \int_a^b f(x)dx$ (why?) and completes the proof of Theorem 3.1.

Example 3.1 For $n \geq 1$, let f_n be defined by

$$f_n(x) = \sqrt{x^2 + \frac{1}{n}}.$$

Let $f(x) = |x|$. We proved in Example 1.4 that (f_n) converges uniformly to f on $\mathbb{R}$. Hence (f_n) converges uniformly to f on $[-1, 1]$ (why?). For every $n \geq 1$, f_n is continuous and therefore Theorem 3.1 applies

$$\lim_{n\to\infty} \int_{-1}^{1} \sqrt{x^2 + \frac{1}{n}} dx = \int_{-1}^{1} |x| dx = 1.$$

4 Uniform Convergence and Differentiation

First, an example.

Example 4.1 Consider again the sequence of functions (f_n) defined on $\mathbb{R}$ by

$$f_n(x) = \sqrt{x^2 + \frac{1}{n}}.$$

We first show that for each $n \geq 1$ the function f_n is differentiable at $x = 0$ (in fact a very similar argument shows that f_n is differentiable everywhere). Let g_n be defined by $g_n(x) = x^2 + \frac{1}{n}$. The function g_n is a polynomial and is therefore differentiable everywhere and hence at $x = 0$. Observe that $g_n(0) = \frac{1}{n}$ and that the function square root is differentiable at $\frac{1}{n}$. In fact the square root function is differentiable at any strictly positive number (but not at 0). Hence, by the chain rule $f_n = \sqrt{g_n}$ is differentiable at $x = 0$.

We have proved in Example 1.4 that the sequence (f_n) converges uniformly to the absolute value function on $\mathbb{R}$. Hence, the limit function is not differentiable at 0! This example shows that uniform convergence does not preserve differentiability!

The next result gives sufficient conditions for the limit function to be differentiable.

Theorem 4.1 *Assume that for all $n \geq 1$ the function f_n is differentiable on $[a, b]$ and the function f_n' is continuous on $[a, b]$.*

Assume that the sequence (f_n) converges pointwise on $[a, b]$ to a function f.

Finally, assume that the sequence of derivatives (f_n') converges uniformly on $[a, b]$ to some function g.

Then the function f is differentiable on $[a, b]$, f' is continuous on $[a, b]$, and $f' = g$.

Note that we do not require (f_n) to converge uniformly. Instead we require (f_n') to converge uniformly.

Proof of Theorem 4.1

Let x and c be in $[a, b]$. For every $n \geq 1$, f_n' is continuous on $[c, x]$ and therefore integrable. By the Fundamental Theorem of Calculus

$$\int_c^x f_n'(t)dt = f_n(x) - f_n(c) \tag{4.1}$$

Since (f_n') converges uniformly to g on $[a, b]$ we have by Theorem 3.1

$$\lim_{n\to\infty} \int_c^x f_n'(t)dt = \int_c^x g(t)dt.$$

On the other hand, since (f_n) converges pointwise to f on (a, b),

$$\lim_{n\to\infty} (f_n(x) - f_n(c)) = f(x) - f(c).$$

Using the two limits above and letting n go to infinity in (4.1) we get

$$\int_c^x g(t)dt = f(x) - f(c) \tag{4.2}$$

Observe now that the functions f_n' are assumed to be continuous and the sequence (f_n') converges uniformly to g. Hence, g is continuous on $[a, b]$. By the second version of the Fundamental Theorem of Calculus, $\int_c^x g(t)dt$ is differentiable as a function of x and

$$\frac{d}{dx} \int_c^x g(t)dt = g(x).$$

Since the l.h.s. of (4.2) is differentiable so is the r.h.s. Hence, $f(x) - f(c)$ is differentiable on $[a, b]$. Hence, f is differentiable on $[a, b]$. By taking derivatives with respect to x on both sides of (4.2) we get

$$g(x) = f'(x).$$

This proves that the sequence (f_n') converges uniformly to f' on $[a, b]$. Moreover, since g is continuous on $[a, b]$ so is f'. This completes the proof of Theorem 4.1.

Problems

1. Assume that (f_n) and (g_n) converge uniformly on some set $S \subset \mathbb{R}$.

(a) Show that the sequence $(f_n + g_n)$ converges uniformly on S.
(b) Show that the sequence $(f_n g_n)$ need not converge uniformly. (Take $f_n(x) = g_n(x) = x + \frac{1}{n}$).
(c) In addition to the uniform convergence of (f_n) and (g_n) assume that these sequences are bounded. That is, assume that there exist $b_1 > 0$ and $b_2 > 0$

such that for all $n \geq 1$ and all x in S we have $|f_n(x)| < b_1$ and $g_n(x) < b_2$. Show that $(f_n g_n)$ converges uniformly on S.

2. For every n, f_n is defined on $[0, 1]$. Let $f_n(0) = 0$ and $f_n(x) = 0$ for $x \in (\frac{1}{n}, 1]$. We have no information on f_n on $(0, \frac{1}{n}]$. Prove that (f_n) converges pointwise to 0 on $[0, 1]$.

3. Prove that uniform convergence implies pointwise convergence.

4. Let

$$f_n(x) = \frac{\sin(nx)}{n}.$$

Show that (f_n) converges uniformly on $\mathbb{R}$.

5. Let

$$f_n(x) = xe^{-nx}.$$

(a) Show that (f_n) converges pointwise to 0 on $[0, \infty)$.
(b) Let $m_n = \sup\{|f_n(x)|; x \in [0, \infty)\}$. Show that $m_n = \frac{1}{ne}$.
(c) Conclude that (f_n) converges uniformly on $[0, \infty)$.

6. Let $f_n(x) = e^{-nx}$ for x in $S = [1, \infty)$ and $n \geq 1$.

(a) Show that (f_n) converges uniformly on S.
(b) Show that for any $a > 0$ (f_n) converges uniformly on $[a, \infty)$.
(c) Does (f_n) converge uniformly on $(0, \infty)$?

7. Consider the sequence of functions (f_n) on $[0, 1]$ defined by $f_n(x) = x^n$.

(a) Show that (f_n) converges pointwise on $[0, 1]$.
(b) Show that (f_n) does not converge uniformly on $[0, 1]$.
(c) Find a subset of $[0, 1]$ on which (f_n) does converge uniformly.

8. Assume that the functions f_n are continuous on $[a, b]$ and that the sequence (f_n) converges uniformly to f on $[a, b]$. Let

$$F_n(x) = \int_a^x f_n(t)dt \text{ and } F(x) = \int_a^x f(t)dt.$$

(a) Show that the functions F_n and F are defined on $[a, b]$.
(b) Prove that (F_n) converges uniformly to F on $[a, b]$.

9. Show that if (f_n) converges uniformly to f on S and $T \subset S$, then (f_n) converges uniformly to f on T as well.

10. Recall that a function f is said to be uniformly continuous on S if for every $\epsilon > 0$ there exists a δ such that if $|x - y| < \delta$, then $|f(x) - f(y)| < \epsilon$. Show that uniform convergence preserves uniform continuity.

11. Let (f_n) be a sequence of functions that are differentiable on $[a, b]$ with continuous derivatives. Assume that there is an x_0 in $[a, b]$ for which the sequence $(f_n(x_0))$ converges. Assume also that the sequence (f'_n) converges uniformly on $[a, b]$ to some g.

(a) Let x be in $[a, b]$. Show that for all $n \geq 1$ and $p \geq 1$ we have

$$\left| f_n(x) - f_p(x)) - ((f_n(x_0) - f_p(x_0)\right| \leq |x - x_0| d(f'_n, f'_p),$$

where $d(f'_n, f'_p) = \max_{y \in [a,b]} |f'_n(y) - f'_p(y)|$ (Apply the Mean Value Theorem to $f_n - f_p$).

(b) Let $\epsilon > 0$. Show that there exists N such that if $n \geq N$ and $p \geq N$, then

$$|f_n(x_0) - f_p(x_0)| < \frac{\epsilon}{2}.$$

(c) Use (a) and (b) to show that (f_n) is uniformly Cauchy on $[a, b]$.
(d) Show that (f_n) converges uniformly on $[a, b]$ to some f.
(e) Show that f is differentiable on $[a, b]$ and $f' = g$.

5 Polynomial Approximation of a Continuous Function

In this section we show that a continuous function is always the uniform limit of a sequence of polynomials.

Theorem 5.1 *Let f be continuous on the closed and bounded interval* $[0, 1]$. *For every $n \geq 1$ let*

$$p_n(x) = \sum_{k=0}^{n} f(\frac{k}{n}) \binom{n}{k} x^k (1 - x)^{n-k}.$$

The sequence (p_n) of polynomials converges uniformly to f on $[0, 1]$.

The polynomials p_n are called **Bernstein polynomials**.

Proof of Theorem 5.1
This is an example of an analysis result proved with probability ideas. See Chapter VII in Feller (1971) for several other examples.

By the binomial theorem,

$$\sum_{k=0}^{n} \binom{n}{k} x^k (1-x)^{n-k} = (x+1-x)^n = 1.$$

Therefore, for x in $[0, 1]$

$$\begin{aligned}
|f(x) - p_n(x)| &= \left| f(x) - \sum_{k=0}^{n} \binom{n}{k} x^k (1-x)^{n-k} f(\frac{k}{n}) \right| \\
&= \left| \sum_{k=0}^{n} \binom{n}{k} x^k (1-x)^{n-k} (f(x) - f(\frac{k}{n})) \right| \\
&\le \sum_{k=0}^{n} \binom{n}{k} x^k (1-x)^{n-k} |f(x) - f(\frac{k}{n})|.
\end{aligned}$$

Since f is continuous on the closed and bounded interval $[0, 1]$, f is uniformly continuous on $[0, 1]$. For a proof see Section 6.3 in Schinazi (2011). In a later chapter we will prove the following more general result. If a function is continuous on a compact set, then it is uniformly continuous on that set.

We now use the definition of uniform continuity. For every $\epsilon > 0$ there exists a $\delta > 0$ such that if $|y - z| < \delta$, then $|f(y) - f(z)| < \frac{\epsilon}{2}$. For a fixed x in $[0, 1]$ and a fixed natural n, let

$$A = \{k \in \mathbb{Z} : 0 \le k \le n \text{ and } |k - nx| < n\delta\}.$$

Note that if k is in A, then $|\frac{k}{n} - x| < \delta$ and $|f(x) - f(\frac{k}{n}))| < \frac{\epsilon}{2}$. Thus,

$$\begin{aligned}
\sum_{k \in A} \binom{n}{k} x^k (1-x)^{n-k} |f(x) - f(\frac{k}{n})| &< \frac{\epsilon}{2} \sum_{k \in A} \binom{n}{k} x^k (1-x)^{n-k} \\
&\le \frac{\epsilon}{2} \sum_{k=0}^{n} \binom{n}{k} x^k (1-x)^{n-k} \\
&= \frac{\epsilon}{2}
\end{aligned}$$

Hence,

$$\sum_{k \in A} \binom{n}{k} x^k (1-x)^{n-k} |f(x) - f(\frac{k}{n})| \le \frac{\epsilon}{2} \tag{5.1}$$

If A is the empty set, we take sums over A to be 0 and (5.1) still holds. We now take care of the k's that are not in A. Let

$$B = \{k \in \mathbb{Z} : 0 \leq k \leq n \text{ and } |k - nx| \geq n\delta\}.$$

A continuous function on a closed bounded interval is bounded. Hence, there exists a real number b such that $|f(x)| \leq b$ for all x in [0, 1]. Therefore,

$$\sum_{k \in B} \binom{n}{k} x^k (1-x)^{n-k} |f(x) - f(\frac{k}{n})| \leq 2b \sum_{k \in B} \binom{n}{k} x^k (1-x)^{n-k} \tag{5.2}$$

To find an upper bound for the r.h.s. we need the following.

Lemma 5.1 *For all x in* [0, 1] *we have*

$$\sum_{k \in B} \binom{n}{k} x^k (1-x)^{n-k} \leq \frac{1}{4n\delta^2}.$$

Proof of Lemma 5.1
We start with

$$\sum_{k=0}^{n} (k - nx)^2 \binom{n}{k} x^k (1-x)^{n-k} \geq \sum_{k \in B} (k - nx)^2 \binom{n}{k} x^k (1-x)^{n-k}.$$

If $k \in B$, then $(k - nx)^2 \geq n^2\delta^2$ and therefore

$$\sum_{k \in B} (k - nx)^2 \binom{n}{k} x^k (1-x)^{n-k} \geq n^2\delta^2 \sum_{k \in B} \binom{n}{k} x^k (1-x)^{n-k}.$$

Hence,

$$\sum_{k \in B} \binom{n}{k} x^k (1-x)^{n-k} \leq \frac{1}{n^2\delta^2} \sum_{k=0}^{n} (k - nx)^2 \binom{n}{k} x^k (1-x)^{n-k}.$$

The sum in the r.h.s. is exactly the variance of a binomial random variable with parameters n and x. It is equal to $nx(1-x)$, see the problems for this computation. Hence,

$$\sum_{k \in B} \binom{n}{k} x^k (1-x)^{n-k} \leq \frac{1}{n^2\delta^2} nx(1-x).$$

Observe now that $x(1-x)$ has a maximum value on [0, 1] at $x = \frac{1}{2}$. Thus, $x(1-x) \leq \frac{1}{4}$ for x in [0, 1]. This completes the proof of Lemma 5.1.

Using Lemma 5.1 in (5.2) yields

$$\sum_{k\in B}\binom{n}{k}x^k(1-x)^{n-k}|f(x)-f(\frac{k}{n})| \leq 2b\frac{1}{4n\delta^2}.$$

Putting together this with (5.1) we get

$$\begin{aligned}|f(x)-p_n(x)| &= \sum_{k\in A}\binom{n}{k}x^k(1-x)^{n-k}|f(x)-f(\frac{k}{n})| \\ &+\sum_{k\in B}\binom{n}{k}x^k(1-x)^{n-k}|f(x)-f(\frac{k}{n})| \\ &\leq \frac{\epsilon}{2}+\frac{b}{2n\delta^2}.\end{aligned}$$

We now pick N such that if $n \geq N$, then

$$\frac{b}{2n\delta^2} < \frac{\epsilon}{2}.$$

Recall that δ does not depend on x (uniform continuity!) and therefore N is uniform in x as well. Thus, for every $\epsilon > 0$ we can find N such that if $n \geq N$ and x is in $[0, 1]$,

$$|f(x)-p_n(x)| < \epsilon.$$

This completes the proof of Theorem 5.1.

Problems

1. Let $f(x) = |x - \frac{1}{2}|$ for x in $[0, 1]$. Compute the first 5 Bernstein polynomials for f. Graph the polynomials and f on the same graph.

2. Show that if f is continuous on $[a, b]$, then there is a sequence of polynomials converging uniformly to f on $[a, b]$. (Consider the function $g(x) = f((b-a)x + a)$.)

3. In this problem we show the formula

$$\sum_{k=0}^{n}(k-nx)^2\binom{n}{k}x^k(1-x)^{n-k} = nx(1-x).$$

(a) Show that

$$\sum_{k=0}^{n} k \binom{n}{k} x^k (1-x)^{n-k} = nx \sum_{k=1}^{n} \binom{n-1}{k-1} x^{k-1} (1-x)^{n-k} = nx.$$

(b) Show that

$$\sum_{k=0}^{n} k(k-1) \binom{n}{k} x^k (1-x)^{n-k} = n(n-1)x^2.$$

(c) Use (a) and (b) to show that

$$\sum_{k=0}^{n} k^2 \binom{n}{k} x^k (1-x)^{n-k} = n(n-1)x^2 + nx.$$

(d) Show that

$$\sum_{k=0}^{n} (k-nx)^2 \binom{n}{k} x^k (1-x)^{n-k} = nx(1-x).$$

4. Let f be a continuous function on $[0, 1]$ such that for every polynomial p we have

$$\int_0^1 f(x)p(x)dx = 0.$$

Show that f is identically 0 on $[0, 1]$.

6 Continuity and Smoothness

A smooth function is a function which is at least differentiable, possibly infinitely differentiable. The absolute value function is an example of a continuous function which is not differentiable at 0 and therefore not smooth. On the other hand, the previous section shows that a continuous function is the uniform limit of polynomials. No function is smoother than a polynomial. So how rough can a continuous function be? Surprisingly, a continuous function can be nowhere differentiable! Even worse, Baire's Category Theorem (see Theorem 12.10 in Krantz (1991)) shows that "most" (in the sense of Baire) continuous functions are nowhere differentiable!

In this section we give an example of a continuous nowhere differentiable function. We first need a uniform convergence test for series of functions.

6.1 M-Weierstrass Test

Theorem 6.1 (M-Weierstrass Test) *Let (f_n) be a sequence of functions defined on $S \subset \mathbb{R}$. Assume that there exists a sequence of real numbers (m_n) such that*

$$\sup_{x \in S} |f_n(x)| < m_n.$$

If the numerical series $\sum_{n\geq 1} m_n$ converges, then the series of functions $\sum_{n\geq 1} f_n$ converges uniformly on S.

Proof of Theorem 6.1
We first show that the sequence of partial sums is uniformly Cauchy. For x in S and $n \geq 1$, let

$$s_n(x) = \sum_{k=1}^{n} f_k(x).$$

Assume $p > n$, by the triangle inequality

$$|s_p(x) - s_n(x)| \leq \sum_{k=n+1}^{p} |f_k(x)|.$$

Since every $|f_k|$ is bounded above by m_k we get for every x in S

$$|s_p(x) - s_n(x)| < \sum_{k=n+1}^{p} m_k.$$

The series $\sum_{n\geq 1} m_n$ converges. We now use the Cauchy criterion for series (see Proposition 3.1 in Chapter 4). For every $\epsilon > 0$, there exists an N such that if $p > n \geq N$, then

$$\sum_{k=n+1}^{p} m_k < \epsilon.$$

Therefore, for $p > n \geq N$ and x in S

$$|s_p(x) - s_n(x)| < \sum_{k=n+1}^{p} m_k < \epsilon.$$

That is (s_n) is uniformly Cauchy. Hence, (s_n) converges uniformly by Theorem 2.2. This completes the proof of Theorem 6.1.

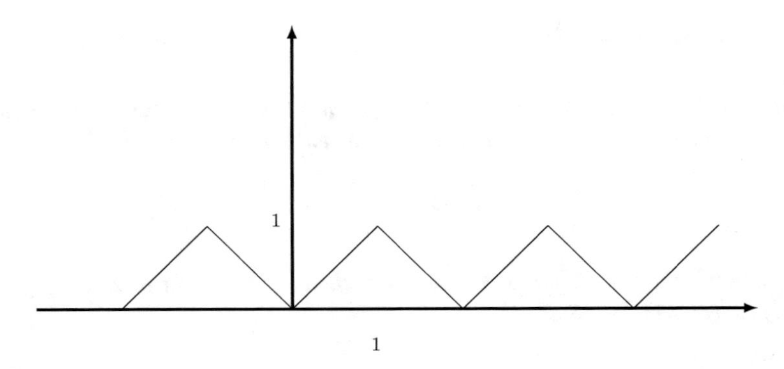

Fig. 5.1 This is the graph of the function Φ.

6.2 *Nowhere Differentiability*

Consider a function Φ such that

$$\Phi(x) = |x| \text{ for } x \in [-1, 1],$$

and such that for every x in $\mathbb{R}$

$$\Phi(x+2) = \Phi(x).$$

That is, Φ has period 2. See Figure 5.1.

Using Φ we define the function f by

$$f(x) = \sum_{k=0}^{\infty} (\frac{3}{4})^k \Phi(4^k x).$$

We now prove that f is continuous everywhere but nowhere differentiable.

- The function f is continuous everywhere on $\mathbb{R}$.

  Observe that for every x, $|\Phi(x)| \leq 1$. Hence, for every x and every $k \geq 0$

$$|\left(\frac{3}{4}\right)^k \Phi(4^k x)| \leq \left(\frac{3}{4}\right)^k.$$

By Theorem 6.1, the sequence (f_n) of partial sums

$$f_n(x) = \sum_{k=0}^{n} (\frac{3}{4})^k \Phi(4^k x),$$

converges uniformly on $\mathbb{R}$. Since for every $n \geq 1$, f_n is continuous on $\mathbb{R}$ so is f (why?).

- The function f is nowhere differentiable.

Let a be in $\mathbb{R}$. We will show that there is a sequence (h_n) converging to 0 such that

$$\frac{f(a+h_n)-f(a)}{h_n}$$

has no limit as n goes to infinity. This is enough to show that f is not differentiable at a. Since we can do this for every a, f is nowhere differentiable!

We now construct the sequence (h_n). Let $n \geq 1$ and consider the open interval

$$(4^n a - \frac{1}{2}, 4^n a + \frac{1}{2}).$$

Since the length of this interval is 1 there is at most one integer in it. If there is an integer in $(4^n a - \frac{1}{2}, 4^n a)$ we set

$$h_n = \frac{1}{2}4^{-n}.$$

If there is an integer in $(4^n a, 4^n a + \frac{1}{2})$ we set

$$h_n = -\frac{1}{2}4^{-n}.$$

If neither interval has an integer we set $h_n = \frac{1}{2}4^{-n}$. Clearly h_n is never 0 and (h_n) converges to 0.

Note that h_n is picked so that there is no integer strictly between $4^n a$ and $4^n a + 4^n h_n$. In the graph of Φ these two points do not have a "corner" between them. Therefore, the points with coordinates $(4^n a, \Phi(4^n a))$ and coordinates $(4^n a + 4^n h_n, \Phi(4^n a + 4^n h_n))$ are on the same line segment. Hence,

$$|\Phi(4^n(a+h_n)) - \Phi(4^n a)| = |4^n(a+h_n) - 4^n a| = |4^n h_n|. \tag{6.1}$$

Observe also that for all x and y we have

$$|\Phi(x) - \Phi(y)| \leq |x-y| \tag{6.2}$$

For fixed a and $n \geq 1$ consider now

$$f(a+h_n) - f(a) = \sum_{k=0}^{\infty} (\frac{3}{4})^k \left(\Phi(4^k(a+h_n)) - \Phi(4^k a)\right).$$

If $k > n$, then

$$|4^k h_n| = \frac{1}{2} 4^{k-n}$$

which is a multiple of 2. Since Φ has period 2 we get for $k > n$

$$\Phi(4^k(a + h_n)) = \Phi(4^k a).$$

Hence,

$$f(a + h_n) - f(a) = \sum_{k=0}^{n} (\frac{3}{4})^k \left(\Phi(4^k(a + h_n)) - \Phi(4^k a) \right).$$

By using the triangle inequality (see Problem 4.) we get

$$|f(a + h_n) - f(a)| \geq (\frac{3}{4})^n \left| \Phi(4^n(a + h_n)) - \Phi(4^n a) \right|$$

$$- \sum_{k=0}^{n-1} (\frac{3}{4})^k |\Phi(4^k(a + h_n)) - \Phi(4^k a)|.$$

Using (6.1) for the term $|\Phi(4^n(a + h_n)) - \Phi(4^n a)|$ and (6.2) for the terms

$$|\Phi(4^k(a + h_n)) - \Phi(4^k a)|,$$

for $k = 0, \ldots, n - 1$, we get

$$|f(a + h_n) - f(a)| \geq (\frac{3}{4})^n 4^n |h_n| - \sum_{k=0}^{n-1} (\frac{3}{4})^k 4^k |h_n|.$$

Dividing both sides by $|h_n|$,

$$\frac{|f(a + h_n) - f(a)|}{|h_n|} \geq 3^n - \sum_{k=0}^{n-1} 3^k = \frac{1}{2}(3^n + 1).$$

This shows that the l.h.s. cannot have a finite limit. This concludes the proof that the function f is not differentiable at a. Since a is arbitrary, f is nowhere differentiable.

- The function Φ that we use is not differentiable at integer points and therefore is not smooth. This is actually not critical. Weierstrass' original example of a continuous nowhere differentiable function is

$$f(x) = \sum_{k=1}^{\infty} b^n \cos\left(a^n \pi x\right),$$

where $b < 1$ and $ab > 1 + \frac{3}{2}\pi$. The cosine function is infinitely differentiable everywhere but f is nowhere differentiable. See Dunham (2005) for a proof.

Problems

1. Let $a = \frac{1}{3}$. Let $n \geq 1$. If there is an integer in $(4^n a - \frac{1}{2}, 4^n a)$ we set

$$h_n = \frac{1}{2} 4^{-n}.$$

If there is an integer in $(4^n a, 4^n a + \frac{1}{2})$ we set

$$h_n = -\frac{1}{2} 4^{-n}.$$

Find h_1, h_2 and h_3.

2. Weierstrass' original example of a nowhere differentiable function is for

$$f(x) = \sum_{k=1}^{\infty} b^n \cos\left(a^n \pi x\right),$$

where $b < 1$ and $ab > 1 + \frac{3}{2}\pi$.

(a) Show for any choice of a and b the function f is continuous on $\mathbb{R}$.
(b) Let

$$f_n(x) = \sum_{k=1}^{n} (\frac{1}{4})^n \cos\left((15)^n \pi x\right).$$

Graph f_1, f_2, and f_3.

3. For $n \geq 0$, let

$$s_n(x) = \sum_{k=0}^{n} c_k x^k.$$

Assume that the sequence $(s_n(x))$ converges for every x in $(-R, R)$ where R is a given real number. Let $0 < r < R$. Use the M-Weierstrass test to prove that (s_n) converges uniformly on $[-r, r]$.

4. Let (a_n) be a sequence of real numbers. Show that for every $n \geq 1$ we have

$$\left|\sum_{k=0}^{n} a_k\right| \geq |a_n| - \sum_{k=0}^{n-1} |a_k|.$$

Chapter 6
Power Series

In this chapter we apply the results on uniform convergence to power series. In particular, we will justify some very useful calculus techniques such as term by term integration and term by term differentiation.

1 Radius of Convergence

A power series is a function

$$f(x) = \sum_{n=0}^{\infty} c_n x^n$$

where (c_n) is a sequence of real numbers. The function f is defined at a if and only if the series converges (pointwise) at $x = a$. By definition, a series converges at $x = a$ if the sequence of partial sums

$$f_n(x) = \sum_{k=0}^{n} c_k x^k$$

converges at $x = a$.

The first task when studying a power series is to decide where it is defined. The (possibly infinite) number

$$R = \frac{1}{\limsup_n |c_n|^{1/n}}$$

is called the **radius of convergence** of the power series $\sum_{n=0}^{\infty} c_n x^n$.

R. B. Schinazi, *From Classical to Modern Analysis*,
https://doi.org/10.1007/978-3-319-94583-5_6

- If $\limsup_n |c_n|^{1/n} = 0$ we set $R = +\infty$.
- If $\limsup_n |c_n|^{1/n} = +\infty$ we set $R = 0$.

Theorem 1.1 *Let R be the radius of convergence of the power series $\sum_{n=0}^{\infty} c_n x^n$. If R is finite the series converges pointwise on $(-R, R)$ and diverges for $|x| > R$. If $R = +\infty$ the power series converges pointwise on $(-\infty, +\infty)$.*

- When R is finite, the series may converge or diverge at $x = R$ and $x = -R$.
- Note that the domain of a power series is always an interval.

Proof of Theorem 1.1
The proof is a simple consequence of the root test. For $n \geq 0$, let

$$a_n = \left|c_n x^n\right|.$$

Then $|a_n|^{1/n} = |a_n|^{1/n}|x|$ and

$$\limsup_n |a_n|^{1/n} = |x| \limsup_n |c_n|^{1/n}.$$

There are three possibilities.

If $\limsup_n |c_n|^{1/n} = 0$, then $\limsup_n |a_n|^{1/n} = 0$ for every x. By the root test the series $\sum_{n=0}^{\infty} c_n x^n$ converges for every x.

If $\limsup_n |c_n|^{1/n} = +\infty$, then $\limsup_n |a_n|^{1/n} = +\infty$ for every $x \neq 0$. By the root test the series $\sum_{n=0}^{\infty} c_n x^n$ diverges for every $x \neq 0$.

If $\limsup_n |c_n|^{1/n}$ is a finite strictly positive real $\bar{\ell}$, then $\limsup_n |a_n|^{1/n} = |x|\bar{\ell}$ for every x. Let $R = 1/\bar{\ell}$. By the root test the series $\sum_{n=0}^{\infty} c_n x^n$ converges for $|x| < R$ and diverges for $|x| > R$.

The proof of Theorem 1.1 is complete.

Example 1.1 What is the domain of the power series $\sum_{n=1}^{\infty} \frac{(-1)^n}{n2^n} x^n$?

Let

$$a_n = \frac{(-1)^n}{n2^n}.$$

Then,

$$|a_n|^{1/n} = \frac{1}{2n^{1/n}}.$$

Since the sequence $(n^{1/n})$ converges to 1 we have that

$$\lim_n |a_n|^{1/n} = \frac{1}{2}.$$

Hence, the radius of convergence for this power series is $R = 2$.

At this point we know that the domain of the series includes $(-2, 2)$ but it may or may not include the end points of this interval. We now examine the endpoints. Plugging $x = 2$ in the series gives

$$\sum_{n=1}^{\infty} \frac{(-1)^n}{n}$$

which converges by the Alternating Series Theorem. On the other hand setting $x = -2$ in the power series gives

$$\sum_{n=1}^{\infty} \frac{1}{n}$$

which diverges (why?). Therefore, the domain of this power series is $(-2, 2]$.

As the next example shows the ratio test can also be useful in finding a radius of convergence.

Example 1.2 What is the radius of convergence of $\sum_{n=0}^{\infty} \frac{x^n}{n!}$?
For a fixed x, let

$$a_n = \frac{|x|^n}{n!}.$$

Thus,

$$\frac{a_{n+1}}{a_n} = \frac{1}{n+1}|x|,$$

which converges to 0 for any x. By the ratio test the series $\sum_{n=0}^{\infty} \frac{x^n}{n!}$ converges for all x. The radius of convergence of this series is therefore $R = +\infty$.

2 Power Series and Uniform Convergence

Let $f(x) = \sum_{n=0}^{\infty} c_n x^n$. For $n \geq 1$ we define

$$f_n(x) = \sum_{k=0}^{n} c_k x^k.$$

Let $R \geq 0$ be the radius of convergence of f. By Theorem 1.1, (f_n) converges pointwise to f on $(-R, R)$. In fact, as we see next we also have uniform convergence inside $(-R, R)$.

Theorem 2.1 *Consider a power series* $f(x) = \sum_{n=0}^{\infty} c_n x^n$ *with a radius of convergence* R *in* $(0, +\infty]$. *For* $n \geq 1$ *and* x *in* $(-R, R)$ *let*

$$f_n(x) = \sum_{k=0}^{n} c_k x^k.$$

Then, for any $0 < r < R$ *the sequence* (f_n) *of partial sums converges uniformly to* f *on* $[-r, r]$.

- We use the notation R in $(0, +\infty]$ to indicate that $R > 0$ and possibly infinite.

Proof of Theorem 2.1
Since $r < R$ we can find r_1 such that $r < r_1 < R$. Assume that $|x| \leq r < r_1$, then

$$|c_k x^k| = |c_k||x|^k = |c_k| r_1^k \frac{|x|^k}{r_1^k} \leq |c_k| r_1^k \frac{r^k}{r_1^k}.$$

Since $-R < r_1 < R$ we know that the series

$$\sum_{k=0}^{\infty} c_k r_1^k$$

converges. Therefore, the sequence $(|c_k| r_1^k)$ is bounded (why?) by some b. Thus,

$$|c_k x^k| \leq b(\frac{r}{r_1})^k.$$

Hence,

$$\sup_{x \in [-r,r]} |c_k x^k| \leq b(\frac{r}{r_1})^k,$$

and the numerical series $\sum_{k \geq 0} b(\frac{r}{r_1})^k$ converges. We can apply the M-Weierstrass test (Theorem 6.1 in Chapter 5) to conclude that the power series $\sum_{k \geq 0} c_k x^k$ converges uniformly on $[-r, r]$. This completes the proof of Theorem 2.1.

Corollary 2.1 *Consider a power series* $f(x) = \sum_{n=0}^{\infty} c_n x^n$ *with a radius of convergence* R *in* $(0, +\infty]$. *Then,* f *is continuous everywhere on* $(-R, R)$.

Proof of Corollary 2.1
Let a be in $(-R, R)$. Then a is in $[-r, r]$ for some $r < R$. Theorem 2.1 shows that the sequence (f_n) of partial sums converges uniformly to f on $[-r, r]$. Note that each f_n is continuous on $[-r, r]$ (why?). Since uniform convergence preserves continuity, f is continuous on $[-r, r]$ and therefore at a. Hence, f is continuous on $(-R, R)$. This completes the proof of Corollary 2.1.

3 Integrating and Differentiating Power Series

The next result shows that term by term integration is legitimate.

Proposition 3.1 *Consider a power series* $f(x) = \sum_{n=0}^{\infty} c_n x^n$ *with a radius of convergence* R *in* $(0, +\infty]$. *Let* $a < b$ *such that* $[a, b] \subset (-R, R)$. *Then* f *is Riemann integrable on* $[a, b]$ *and*

$$\int_a^b f(x)dx = \sum_{n=0}^{\infty} c_n \int_a^b x^n dx.$$

Proof of Proposition 3.1
First note that there is r in $(0, R)$ such that $[a, b] \subset [-r, r]$. Hence, the sequence of partial sums (f_n) converges uniformly on $[-r, r]$ and therefore on $[a, b]$. Observe also that for any $n \geq 0$, f_n is a polynomial and is therefore continuous. Hence, by Theorem 3.1 in Chapter 5

$$\lim_{n\to\infty} \int_a^b f_n(x)dx = \int_a^b \lim_{n\to\infty} f_n(x) = \int_a^b f(x)dx \tag{3.1}$$

By definition of f_n we have

$$\int_a^b f_n(x)dx = \int_a^b (\sum_{k=0}^{n} c_k x^k)dx = \sum_{k=0}^{n} c_k \int_a^b x^k dx.$$

The critical point here is that we can always interchange the integral and the sum if the sum has **finitely** many terms (provided the functions are integrable). To do so we just use the linearity property of the Riemann integral. However, to do the interchange for an infinite sum we need more. Here we will use uniform convergence.

Back to the proof, the last equality shows that $\int_a^b f_n(x)dx$ is the partial sum of the series with general term $c_k \int_a^b x^k dx$. By (3.1) we know that this partial sum converges to $\int_a^b f(x)dx$. Hence,

$$\lim_{n\to\infty} \sum_{k=0}^{n} c_k \int_a^b x^k dx = \int_a^b f(x)dx.$$

Since

$$\lim_{n\to\infty} \sum_{k=0}^{n} c_k \int_a^b x^k dx = \sum_{k=0}^{\infty} c_k \int_a^b x^k dx,$$

we have

$$\sum_{k=0}^{\infty} c_k \int_a^b x^k dx = \int_a^b f(x)dx.$$

This completes the proof of Proposition 3.1.

An important application of the preceding result is to get new power series from known ones. The next example illustrates this.

Example 3.1 Recall from calculus (see for instance Section 4.1 in Schinazi (2011)) the binomial series

$$\frac{1}{\sqrt{1-x^2}} = 1 + \sum_{k=1}^{\infty} \frac{(2k)!}{2^{2k}(k!)^2} x^{2k}.$$

The ratio test shows that this power series has a radius of convergence equal to 1. Let y be in $(0, 1)$, then $[0, y]$ is a subset of $(-1, 1)$ and by Proposition 3.1 we can integrate the series term by term:

$$\int_0^y \frac{1}{\sqrt{1-x^2}} dx = \int_0^y 1 dx + \sum_{k=1}^{\infty} \frac{(2k)!}{2^{2k}(k!)^2} \int_0^y x^{2k} dx.$$

Now, an antiderivative of $\frac{1}{\sqrt{1-x^2}}$ is $\arcsin x$. Hence, by the Fundamental Theorem of Calculus

$$\arcsin y - \arcsin 0 = y + \sum_{k=1}^{\infty} \frac{(2k)!}{2^{2k}(k!)^2} \frac{y^{2k+1}}{2k+1}$$

Since $\arcsin 0 = 0$, we have

$$\arcsin y = y + \sum_{k=1}^{\infty} \frac{(2k)!}{2^{2k}(k!)^2} \frac{y^{2k+1}}{2k+1}$$

for all y in $(0, 1)$. In fact, the same argument works for any y in $(-1, 1)$.

Proposition 3.2 *Consider a power series $f(x) = \sum_{n=0}^{\infty} c_n x^n$ with a strictly positive radius of convergence R. Then f is differentiable on $(-R, R)$. Moreover, for x in $(-R, R)$ we have*

$$f'(x) = \sum_{n=1}^{\infty} n c_n x^{n-1}.$$

In words, a power series can be differentiated term by term.

Proof of Proposition 3.2
For $n \geq 1$ and x in $(-R, R)$ let

$$g_n(x) = f_n'(x) = \sum_{k=1}^{n} kc_k x^{k-1},$$

where f_n is the partial sum of the power series $\sum_{n=0}^{\infty} c_n x^n$. Let $0 < r < R$. Our first step is to prove that the sequence (g_n) converges uniformly to

$$g(x) = \sum_{n=1}^{\infty} nc_n x^{n-1}$$

on $[-r, r]$. Let $r < r_1 < R$, we have for $k \geq 1$

$$|kc_k x^{k-1}| = k|c_k|r_1^{k-1}(\frac{|x|}{r_1})^{k-1}.$$

Since $-R < r_1 < R$ the series $\sum_{n=0}^{\infty} c_n r_1^n$ converges. Thus, the sequence $(|c_k|r_1^{k-1})$ is bounded by some b. Hence,

$$|kc_k x^{k-1}| \leq bk(\frac{|x|}{r_1})^{k-1} < bk(\frac{r}{r_1})^{k-1}$$

for $|x| < r$. It is easy to check that the numerical series

$$\sum_{k=1}^{\infty} bk(\frac{r}{r_1})^{k-1}$$

converges. Hence, by the M-Weierstrass test (Theorem 6.1 in Chapter 5) we have that the sequence (g_n) converges uniformly to g on $[-r, r]$.

We now check that the two other hypotheses of Theorem 4.1 in Chapter 5 hold. First, for any $n \geq 1$

$$f_n(x) = \sum_{k=1}^{n} c_k x^k$$

is infinitely differentiable on $[-r, r]$. Second, by Theorem 2.1 the sequence (f_n) converges uniformly (and therefore pointwise) to f on $[-r, r]$. These facts together with the uniform convergence on $[-r, r]$ of (f_n') imply that f is differentiable on $[-r, r]$. Moreover, $f' = g$ on $[-r, r]$ for any $r < R$. This completes the proof of Proposition 3.2.

An easy consequence of Proposition 4.1 is the following.

Corollary 3.1 *Consider a power series* $f(x) = \sum_{n=0}^{\infty} c_n x^n$ *with a radius of convergence* R *in* $(0, +\infty]$. *For any* $k \geq 1$ *the function* f *can be differentiated* k *times and its* k*-th derivative* $f^{(k)}$ *is given by,*

$$f^{(k)}(x) = \sum_{n=k}^{\infty} n(n-1)\dots(n-k+1)c_n x^{n-k},$$

for any x *in* $(-R, R)$. *In particular,*

$$f^{(k)}(0) = k!c_k.$$

Using Proposition 3.2, Corollary 3.1 can be proved by induction on k. The details are left to the problems.

Proposition 3.3 *Assume that the power series* $\sum_{k\geq 0} a_k x^k$ *has a radius of convergence* R. *Then the power series*

$$\sum_{k\geq 1} k a_k x^{k-1} \textit{ and } \sum_{k\geq 0} \frac{a_k}{k+1} x^{k+1}$$

have both the same radius of convergence R.

In words, taking the derivative or the antiderivative of a power series does not change the radius of convergence.

Proof of Proposition 3.3
Since $|na_n| \geq |a_n|$ for every $n \geq 1$ we have

$$\limsup_n |na_n|^{1/n} \geq \limsup_n |a_n|^{1/n}. \tag{3.2}$$

Since the radius of convergence of a series is defined as the inverse of the corresponding limit superior, (3.2) implies that $R_1 \leq R$ where R_1 is the radius of convergence of $\sum_{k\geq 1} k a_k x^{k-1}$. On the other hand, by Proposition 3.2 the series $\sum_{k\geq 1} k a_k x^{k-1}$ converges on $(-R, R)$. Hence, $R_1 \geq R$. This proves that $R_1 = R$.

We now turn to the antiderivative

$$g(x) = \sum_{k\geq 0} \frac{a_k}{k+1} x^{k+1}.$$

We just proved that a power series has the same radius of convergence as its derivative. Since

$$g'(x) = \sum_{k\geq 0} a_k x^k,$$

g has a radius of convergence equal to R. The proof of Proposition 3.3 is complete.

Corollary 3.2 *Let $f(t) = \sum_{k\geq 0} a_k t^k$ and $g(t) = \sum_{k\geq 0} b_k t^k$. Let $\delta > 0$. Assume that $f(t)$ and $g(t)$ converge for all t in $(-\delta, \delta)$. Assume also that $f(t) = g(t)$ for all t in $(-\delta, \delta)$. Then, $a_k = b_k$ for all $k \geq 0$.*

Proof of Corollary 3.2
Note that $f(0) = a_0$ and $g(0) = b_0$ and therefore $a_0 = b_0$. By Corollary 3.1, the functions f and g are infinitely differentiable at 0 and for all $k \geq 1$, $f^{(k)}(0) = k!a_k$ and $g^{(k)}(0) = k!b_k$. Since $f = g$ on $(-\delta, \delta)$ we have $f^{(k)}(0) = g^{(k)}(0)$ for all $k \geq 1$. Thus, $a_k = b_k$ for all $k \geq 1$. This completes the proof of Corollary 3.2.

4 Convergence at Endpoints

Theorem 4.1 (Abel's Theorem) *Assume that $\sum_{n\geq 1} c_n$ converges. Then the power series $\sum_{n\geq 1} c_n x^n$ converges uniformly on $[0, 1]$.*

Knowing that uniform convergence holds on $[0, 1]$ (and not only on $[0, r]$ for any $r < 1$) turns out to be important in several applications. We will give an example below.

Proof of Theorem 4.1
For x in $[0, 1]$ and $n \geq 1$ let

$$s_n(x) = \sum_{k=1}^{n} c_k x^k.$$

We will show that the sequence (s_n) is uniformly Cauchy on $[0, 1]$.

For $k \geq 1$ let

$$q_k = \sum_{j=k}^{\infty} c_j = c_k + c_{k+1} + \dots.$$

Since $\sum_{j\geq 1} c_j$ converges so does $\sum_{j=k}^{\infty} c_j$ for every k (why?). Hence, q_k is a finite number for every $k \geq 1$.

Note that for all $k \geq 1$ we have

$$c_k = q_k - q_{k+1}.$$

Let $p > n$ and x in $[0, 1]$, then

$$s_p(x) - s_n(x) = \sum_{k=n+1}^{p} c_k x^k = \sum_{k=n+1}^{p} (q_k - q_{k+1})x^k.$$

Therefore,

$$\begin{aligned} s_p(x) - s_n(x) = &\sum_{k=n+1}^{p} q_k x^k - \sum_{k=n+1}^{p} q_{k+1} x^k \\ =& q_{n+1}x^{n+1} + q_{n+2}x^{n+2} + \cdots + q_p x^p \\ &- (q_{n+2}x^{n+1} + \cdots + q_p x^{p-1} + q_{p+1}x^p) \\ =& q_{n+1}x^{n+1} + q_{n+2}(x^{n+2} - x^{n+1}) \\ &+ \cdots + q_p(x^p - x^{p-1}) - q_{p+1}x^p \\ =& q_{n+1}x^{n+1} + (x-1)\left(q_{n+2}x^{n+1} + \cdots + q_p x^{p-1}\right) - q_{p+1}x^p \end{aligned}$$

To complete the proof that the sequence (s_n) is uniformly Cauchy we will show that the three terms on the r.h.s. of the last equality can be made small.

Since $\sum_{j=1}^{\infty} c_j$ converges, as k goes to infinity

$$\sum_{j=1}^{\infty} c_j - \sum_{j=1}^{k} c_j$$

converges to 0. That is, the sequence (q_k) converges to 0. Therefore, there exists N such that if $n \geq N$, then $|q_n| < \frac{\epsilon}{3}$. For all x in $[0, 1]$ and $n \geq N$ we have

$$|q_{n+1}x^{n+1}| \leq |q_{n+1}| < \frac{\epsilon}{3}.$$

Similarly, for $p > n \geq N$ we have

$$|-q_{p+1}x^p| < \frac{\epsilon}{3}.$$

Finally, we have for all x in $[0, 1]$

$$\begin{aligned} |(x-1)\left(q_{n+2}x^{n+1} + \cdots + q_p x^{p-1}\right)| \leq& (1-x)\left(|q_{n+2}|x^{n+1} + \cdots + |q_p|x^{p-1}\right) \\ <& (1-x)\frac{\epsilon}{3}\left(x^{n+1} + \cdots + x^{p-1}\right). \end{aligned}$$

Note that for all x in $[0, 1]$ we have

$$(1-x)\frac{\epsilon}{3}\left(x^{n+1} + \cdots + x^{p-1}\right) \leq (1-x)\frac{\epsilon}{3}\sum_{k=n+1}^{\infty} x^k = \frac{\epsilon}{3}x^{n+1} \leq \frac{\epsilon}{3}.$$

Putting together our three estimates we get for $p > n \geq N$ and x in $[0, 1]$

$$|s_p(x) - s_n(x)| < \frac{\epsilon}{3} + \frac{\epsilon}{3} + \frac{\epsilon}{3} = \epsilon.$$

Hence, (s_n) is uniformly Cauchy on $[0, 1]$. By Theorem 2.2 in Chapter 5, (s_n) converges uniformly on $[0, 1]$. The proof of Theorem 4.1 is complete.

Example 4.1 Show that

$$\arcsin y = y + \sum_{k=1}^{\infty} \frac{(2k)!}{2^{2k}(k!)^2} \frac{y^{2k+1}}{2k+1}$$

for all y in $[0, 1]$ and that the convergence of the series above is uniform on $[0, 1]$.

From Example 3.1 we know that the formula above holds for all y in $(-1, 1)$. We show first that the series above converges at $y = 1$. We use Stirling's formula, see for instance Schinazi (2011)

$$n! \sim \sqrt{2\pi} e^{-n} n^{n+1/2}$$

where $a_n \sim b_n$ means that the sequence $(\frac{a_n}{b_n})$ converges to 1 as n goes to infinity. From Stirling's formula and operations on limits we get

$$\frac{(2k)!}{2^{2k}(k!)^2} \sim \frac{\sqrt{2\pi} e^{-2k}(2k)^{2k+1/2}}{(\sqrt{2\pi} e^{-k} k^{k+1/2})^2}.$$

After cancellations we have

$$\frac{(2k)!}{2^{2k}(k!)^2} \sim \frac{1}{\sqrt{2\pi}} \frac{2^{2k+1/2}}{\sqrt{k}}.$$

Hence,

$$\frac{(2k)!}{2^{2k}(k!)^2} \frac{1}{2k+1} \sim \frac{1}{2\sqrt{\pi} k^{3/2}}.$$

By the p test ($p = 3/2 > 1$) we have that the series

$$\sum_{k \geq 1} \frac{1}{2\sqrt{\pi} k^{3/2}} \text{ converges.}$$

By the limit comparison test the series

$$\sum_{k \geq 1} \frac{(2k)!}{2^{2k}(k!)^2} \frac{1}{2k+1}$$

converges as well. Therefore, by Theorem 4.1 the series

$$\sum_{k=1}^{\infty} \frac{(2k)!}{2^{2k}(k!)^2} \frac{y^{2k+1}}{2k+1}$$

converges uniformly on [0, 1]. Hence, this power series is continuous on [0, 1]. Since

$$\arcsin y = y + \sum_{k=1}^{\infty} \frac{(2k)!}{2^{2k}(k!)^2} \frac{y^{2k+1}}{2k+1}$$

for y in [0, 1) and both sides are continuous at $y = 1$ the equality also holds for $y = 1$ (why?). Hence, the formula holds for all y in [0, 1].

5 Euler's Formula

In this section we prove Euler's formula. We use Stirling's formula and Wallis' integrals from calculus. See for instance Section 4.2 in Schinazi (2011).

Theorem 5.1 (Euler's Formula)

$$\sum_{k=1}^{\infty} \frac{1}{k^2} = \frac{\pi^2}{6}.$$

Proof of Theorem 5.1
We follow the approach of Choe (1987).

We first prove that

$$\int_0^{\frac{\pi}{2}} t dt = \int_0^{\frac{\pi}{2}} \sin t dt + \sum_{n=1}^{\infty} \frac{(2n)!}{2^{2n}(n!)^2(2n+1)} \int_0^{\frac{\pi}{2}} \sin^{2n+1} t dt \tag{5.1}$$

For $n \geq 1$ and y in [0, 1] let

$$f_n(y) = y + \sum_{k=1}^{n} \frac{(2k)!}{2^{2k}(k!)^2} \frac{y^{2k+1}}{2k+1}.$$

For t in $[0, \frac{\pi}{2}]$ let

$$g_n(t) = f_n(\sin t).$$

By Example 4.1 we know that (f_n) converges uniformly to arcsin on [0, 1]. Hence, (g_n) converges uniformly to g on $[0, \frac{\pi}{2}]$, where $g(t) = t$ (why?). By Theorem 3.1 in Chapter 5,

$$\lim_{n\to\infty} \int_0^{\frac{\pi}{2}} g_n(t)dt = \int_0^{\frac{\pi}{2}} g(t)dt.$$

Since,

$$g_n(t) = \sin t + \sum_{k=1}^{n} \frac{(2k)!}{2^{2k}(k!)^2} \frac{\sin^{2k+1} t}{2k+1},$$

we have

$$\int_0^{\frac{\pi}{2}} g_n(t)dt = \int_0^{\frac{\pi}{2}} \sin t dt + \sum_{k=1}^{n} \frac{(2k)!}{2^{2k}(k!)^2} \int_0^{\frac{\pi}{2}} \frac{\sin^{2k+1} t}{2k+1} dt.$$

and therefore,

$$\begin{aligned} \lim_{n\to\infty} \int_0^{\frac{\pi}{2}} g_n(t)dt &= \int_0^{\frac{\pi}{2}} \sin t dt + \sum_{n=1}^{\infty} \frac{(2n)!}{2^{2n}(n!)^2(2n+1)} \int_0^{\frac{\pi}{2}} \sin^{2n+1} t dt \\ &= \int_0^{\frac{\pi}{2}} g(t)dt \\ &= \int_0^{\frac{\pi}{2}} t dt \end{aligned}$$

This completes the proof of formula (5.1).

We recognize Wallis' integrals in (5.1)

$$I_{2n+1} = \int_0^{\pi/2} \sin^{2n+1} t.$$

From calculus we know that these integrals can be computed by induction

$$I_{2n+1} = \frac{2.4.6.\ldots 2n}{3.5.7.\ldots(2n+1)}.$$

Note that

$$2.4.6.\ldots.(2n) = 2^n(1.2.3.\ldots.n) = 2^n n!. \tag{5.2}$$

Observe also that

$$1.3.5.\dots(2n+1)\times 2.4.6\dots(2n)=(2n+1)!. \tag{5.3}$$

By multiplying the numerator and denominator by $2.4.6.\dots.(2n)$ we get

$$I_{2n+1}=\frac{(2.4.6.\dots 2n)^2}{(2.4.\dots 2n)3.5.7.\dots(2n+1)}=\frac{(2^n n!)^2}{(2n+1)!}$$

where we used (5.2) in the numerator and (5.3) in the denominator. Thus,

$$\int_0^{\frac{\pi}{2}}\sin^{2n+1}t\,dt=\frac{(2^n n!)^2}{(2n+1)!}.$$

Using the last computation in (5.1),

$$\int_0^{\pi/2}t\,dt=\int_0^{\pi/2}\sin t\,dt+\sum_{n=1}^{\infty}\frac{(2n)!}{2^{2n}(n!)^2(2n+1)}\frac{(2^n n!)^2}{(2n+1)!} \tag{5.4}$$

Note that

$$\int_0^{\pi/2}t\,dt=t^2/2\Big]_0^{\pi/2}=\pi^2/8,$$

$$\int_0^{\pi/2}\sin t\,dt=-\cos t\Big]_0^{\pi/2}=1,$$

and

$$\frac{(2n)!}{2^{2n}(n!)^2(2n+1)}\frac{(2^n n!)^2}{(2n+1)!}=\frac{1}{(2n+1)^2}.$$

Using these three computations in (5.4) we have

$$\frac{\pi^2}{8}=1+\sum_{n=1}^{\infty}\frac{1}{(2n+1)^2}.$$

Observe that

$$\sum_{n=1}^{\infty}\frac{1}{n^2}=1+\sum_{n=1}^{\infty}\frac{1}{(2n+1)^2}+\sum_{n=1}^{\infty}\frac{1}{(2n)^2}.$$

But

$$\sum_{n=1}^{\infty} \frac{1}{(2n)^2} = \frac{1}{4} \sum_{n=1}^{\infty} \frac{1}{n^2}.$$

Hence,

$$\sum_{n=1}^{\infty} \frac{1}{n^2} = \frac{\pi^2}{8} + \frac{1}{4} \sum_{n=1}^{\infty} \frac{1}{n^2}.$$

Solving for $\sum_{n=1}^{\infty} \frac{1}{n^2}$ yields

$$\sum_{n=1}^{\infty} \frac{1}{n^2} = \frac{\pi^2}{6}.$$

This completes the proof of Euler's formula.

Problems

1. What is the radius of convergence of the series $\sum_{n=0}^{\infty} n!x^n$

2. What is the domain of the series $\sum_{n=1}^{\infty} \frac{1}{n\sqrt{n}3^n} x^n$?

3. What is the domain of the series $\sum_{n=0}^{\infty} \frac{n}{n+1} x^n$?

4. Assume that $c_n \neq 0$ for all $n \geq 1$. Assume that the sequence

$$\left(\frac{|c_{n+1}|}{|c_n|}\right) \text{ converges to a finite limit } \ell.$$

(a) Show that if $\ell = 0$, then the radius of convergence of the series $\sum_{n=0}^{\infty} c_n x^n$ is infinite.

(b) Show that if $\ell > 0$, then the radius of convergence of the series $\sum_{n=0}^{\infty} c_n x^n$ is $1/\ell$.

5. Assume that the sequence $(|c_{n+1}|/|c_n|)$ goes to infinity. Show that the power series $f(x) = \sum_{n=0}^{\infty} c_n x^n$ is defined at $x = 0$ only.

6. Assume the sequence $(c_n b^n)$ is bounded for some $b \neq 0$. Show that the power series $\sum_{n=0}^{\infty} c_n x^n$ converges absolutely for any x such that $|x| < |b|$.

7. Let (c_n) be a sequence of real numbers and let

$$A = \{r \geq 0 : \text{ the sequence } (c_n r^n) \text{ is bounded}\}.$$

We define $\rho = \sup A$ if A is bounded above and $\rho = +\infty$ otherwise. Show that ρ is the radius of convergence of the series $\sum_{n=0}^{\infty} c_n x^n$. (Use Problem 6.).

8. Prove (by induction on k) Corollary 3.1.

9. Assume that f and g are continuous on $[0, 1]$ and that $f = g$ on $[0, 1)$. Show that $f(1) = g(1)$.

10. Assume that (f_n) converges uniformly to f on $[0, 1]$. Let $g_n(t) = f_n(\sin t)$. Show that (g_n) converges uniformly on $[0, \frac{\pi}{2}]$.

11. Starting with the geometric series $\sum_{n\geq 0} x^n = \frac{1}{1-x}$ for $|x| < 1$ this problem derives a famous formula for π.

(a) Show that for y in $(-1, 1)$ we have

$$\frac{1}{1+y^2} = \sum_{n=0}^{\infty}(-1)^n y^{2n}.$$

(b) Show that for y in $[0, 1)$ we have

$$\arctan y = \sum_{n=0}^{\infty}(-1)^n \frac{y^{2n+1}}{2n+1}.$$

(c) Show that in fact the formula above holds for $y = 1$ as well.
(d) Show that

$$\frac{\pi}{4} = \sum_{n=0}^{\infty}(-1)^n \frac{1}{2n+1}.$$

12. Prove that for every integer $n \geq 0$ we have

$$\int_0^{\pi/2} \sin^{2n+1} t = I_{2n+1} = \frac{2.4.6.\ldots.2n}{3.5.7.\ldots.(2n+1)}.$$

13. Let (c_n) be a sequence of positive real numbers. Let $f(x) = \sum_{n=0}^{\infty} c_n x^n$.

(a) Show the existence of the (possibly infinite) left limit

$$\lim_{x\to 1^-} f(x) = \ell.$$

(b) Show that

$$\sum_{n=0}^{\infty} c_n = \ell.$$

Chapter 7
Metric Spaces

1 Definition and Examples

In this chapter we define metric spaces. This will give us a more abstract point of view as compared to the preceding chapters. It will enable us to distinguish between properties that hold specifically for certain sets (for the set of real numbers or for the set of continuous functions, for instance) from the properties that hold in any metric space.

Definition 1.1 *A metric space* (X, d) *is a nonempty set* X *and a real valued function* d *on* $X \times X$ *with the following properties.*

(i) $d(x, y) \geq 0$ *for all* x *and* y *in* X.
(ii) $d(x, y) = 0$ *if and only if* $x = y$.
(iii) $d(x, y) = d(y, x)$ *for all* x *and* y *in* X.
(iv) $d(x, z) \leq d(x, y) + d(y, z)$ *for all* x, y, *and* z *in* X.

Property (iv) is called the **triangle inequality** for the metric d.

In our first example we check that the set of real numbers equipped with the Euclidean metric is indeed a metric space.

1.1 Euclidean Spaces

Example 1.1 Let $X = \mathbb{R}$ and $d(x, y) = |x - y|$ for any x and y in X. It is easy to check that (i) and (iii) hold and we leave this to the reader. For (ii) assume that $d(x, y) = |x - y| = 0$. Note that the absolute value of a number is 0 if and only if the number is 0. Hence, $x - y = 0$. That is, $x = y$ and (ii) holds.

R. B. Schinazi, *From Classical to Modern Analysis*,
https://doi.org/10.1007/978-3-319-94583-5_7

For the triangle inequality (iv) we simply write

$$d(x,z) = |x-z| = |x-y+y-z| \leq |x-y| + |y-z| = d(x,y) + d(y,z).$$

Hence, d is a metric on $\mathbb{R}$. It is the so-called **Euclidean** metric.

Notation. We will use the notation $(\mathbb{R}, |.|)$ for the metric space of the real numbers equipped with the Euclidean metric.

In order to show that the Euclidean metric in higher dimensions is indeed a metric we need two inequalities.

Proposition 1.1 (Cauchy-Schwarz Inequality) *For any natural n, real numbers $a_1, a_2 \dots a_n$ and $b_1, b_2 \dots b_n$ we have*

$$\left(\sum_{i=1}^{n} a_i b_i\right)^2 \leq \sum_{i=1}^{n} a_i^2 \sum_{i=1}^{n} b_i^2.$$

Proof of Proposition 1.1
Let p be the quadratic polynomial defined by

$$p(t) = \sum_{i=1}^{n} (a_i + tb_i)^2$$

for every t in $\mathbb{R}$. Expand the squares to get

$$p(t) = \sum_{i=1}^{n} a_i^2 + 2t \sum_{i=1}^{n} a_i b_i + t^2 \sum_{i=1}^{n} b_i^2.$$

Since $p(t) \geq 0$ for all t we know that the equation $P(t) = 0$ has at most one solution (why?). From algebra we know that an equation

$$at^2 + 2bt + c = 0$$

has at most one solution if and only if $b^2 \leq ac$. Hence,

$$\left(\sum_{i=1}^{n} a_i b_i\right)^2 \leq \sum_{i=1}^{n} a_i^2 \sum_{i=1}^{n} b_i^2.$$

This completes the proof of Proposition 1.1.

Another useful inequality is the following.

Proposition 1.2 (Minkowski Inequality) *For any natural n, real numbers $a_1, a_2 \dots a_n$ and $b_1, b_2 \dots b_n$ we have*

$$\sqrt{\sum_{i=1}^{n}(a_i+b_i)^2} \le \sqrt{\sum_{i=1}^{n} a_i^2} + \sqrt{\sum_{i=1}^{n} b_i^2}.$$

Proof of Proposition 1.2
We have

$$\sum_{i=1}^{n}(a_i+b_i)^2 = \sum_{i=1}^{n} a_i^2 + 2\sum_{i=1}^{n} a_i b_i + \sum_{i=1}^{n} b_i^2.$$

By Proposition 1.1 we have (recall that $\sqrt{x^2} = |x|$)

$$\sqrt{\left(\sum_{i=1}^{n} a_i b_i\right)^2} = \left|\sum_{i=1}^{n} a_i b_i\right| \le \sqrt{\sum_{i=1}^{n} a_i^2}\sqrt{\sum_{i=1}^{n} b_i^2}.$$

Hence,

$$\sum_{i=1}^{n}(a_i+b_i)^2 \le \sum_{i=1}^{n} a_i^2 + 2\sqrt{\sum_{i=1}^{n} a_i^2}\sqrt{\sum_{i=1}^{n} b_i^2} + \sum_{i=1}^{n} b_i^2 = \left(\sqrt{\sum_{i=1}^{n} a_i^2} + \sqrt{\sum_{i=1}^{n} b_i^2}\right)^2.$$

By taking square roots on both sides of the inequality we get

$$\sqrt{\sum_{i=1}^{n}(a_i+b_i)^2} \le \sqrt{\sum_{i=1}^{n} a_i^2} + \sqrt{\sum_{i=1}^{n} b_i^2}.$$

This completes the proof of Proposition 1.2.

The **Euclidean metric** on $\mathbb{R}^n$ is defined as follows. Let $\mathbf{a} = (a_1, \dots, a_n)$ and $\mathbf{b} = (b_1, \dots, b_n)$ be in $\mathbb{R}^n$, then

$$d(\mathbf{a}, \mathbf{b}) = \sqrt{\sum_{i=1}^{n}(a_i - b_i)^2}.$$

The reader can easily check that (i), (ii), and (iii) of definition 1.1 hold. For the triangle inequality (iv) we have for $\mathbf{a}$, $\mathbf{b}$ and $\mathbf{c}$ in $\mathbb{R}^n$

$$d(\mathbf{a}, \mathbf{b}) = \sqrt{\sum_{i=1}^{n}(a_i - b_i)^2} = \sqrt{\sum_{i=1}^{n}(a_i - c_i + c_i - b_i)^2}.$$

We now use Minkowski inequality to get

$$d(\mathbf{a}, \mathbf{b}) \leq \sqrt{\sum_{i=1}^{n}(a_i - c_i)^2} + \sqrt{\sum_{i=1}^{n}(c_i - b_i)^2} = d(\mathbf{a}, \mathbf{c}) + d(\mathbf{c}, \mathbf{b}).$$

Hence, (iv) holds. That is, d is indeed a metric on $\mathbb{R}^n$.

We now turn to metrics on function spaces.

1.2 Uniform Metric on Continuous Functions

Let $C([a, b])$ be the set of real-valued functions that are continuous on the closed and bounded interval $[a, b]$. Let f and g be in $C([a, b])$. We will show that

$$d(f, g) = \max_{x \in [a,b]} |f(x) - g(x)|$$

is a metric on $C([a, b])$.

First we need to check that $d(f, g)$ is defined for all f and g in $C([a, b])$. In other words, why does the function $|f - g|$ have a maximum on $[a, b]$? Since f and g are continuous so is $|f - g|$ (why?). By the Extreme Value Theorem a continuous function on a closed bounded interval $[a, b]$ reaches its maximum and minimum values. Hence, the function $|f - g|$ does have a maximum on $[a, b]$ and $d(f, g)$ is therefore well defined on $C([a, b])$. We now check that d is a metric.

Since $|f - g|$ is a positive function its maximum is positive and so $d(f, g) \geq 0$. Assume that $d(f, g) = 0$. For every x in $[a, b]$ we have

$$0 \leq |f(x) - g(x)| \leq d(f, g) = 0.$$

Therefore, $f(x) = g(x)$ for all x in $[a, b]$. That is, $f = g$ on $[a, b]$. Conversely, if $f = g$, then $d(f, g) = 0$. This proves (ii).

Property (iii) is clear.

For (iv) let f, g, and h be continuous on $[a, b]$. We have for every x in $[a, b]$ that

$$\begin{aligned}|f(x) - g(x)| &= |f(x) - h(x) + h(x) - g(x)| \\ &\leq |f(x) - h(x)| + |h(x) - g(x)| \\ &\leq d(f, h) + d(h, g).\end{aligned}$$

This shows that $d(f, h)+d(h, g)$ is an upper bound of $\{|f(x)-g(x)|; x \in [a, b]\}$. An upper bound is larger than or equal to the maximum of the set. Hence,

$$\max_{x \in [a,b]} |f(x) - g(x)| \leq d(f, h) + d(h, g).$$

That is,

$$d(f, g) \leq d(f, h) + d(h, g).$$

This proves the triangle inequality and therefore that d is a metric on $C([a, b])$.

1.3 Riemann Integral Metric on Continuous Functions

Let f and g be in the space of continuous functions $C([a, b])$. The Riemann integral metric is defined by

$$d(f, g) = \int_a^b |f(x) - g(x)|dx.$$

Note that since f and g are continuous on $[a, b]$ so is $|f - g|$. Hence, $|f - g|$ is Riemann integrable on $[a, b]$ and $d(f, g)$ is defined for all continuous functions f and g.

In order to show that d is a metric we need several elementary properties of the Riemann integral from calculus, see for instance Schinazi (2011).

- Let $a < b$ and assume that f and g are Riemann integrable on $[a, b]$. Let c_1 and c_2 be two real numbers. Then, $c_1 f + c_2 g$ is Riemann integrable on $[a, b]$ and

$$\int_a^b (c_1 f + c_2 g)(x)dx = c_1 \int_a^b f(x)dx + c_2 \int_a^b g(x)dx.$$

- Assume that f is Riemann integrable on $[a, b]$. Let $a < c < b$, then f is Riemann integrable on $[a, c]$ and on $[c, b]$. Moreover,

$$\int_a^b f(x)dx = \int_a^c f(x)dx + \int_c^b f(x)dx.$$

- Let $a < b$ and assume that f and g are Riemann integrable on $[a, b]$. Assume also that

$$f(x) \leq g(x) \text{ for all } x \in [a, b].$$

Then,

$$\int_a^b f(x)dx \leq \int_a^b g(x)dx.$$

- Assume that f is bounded and is continuous on $[a, b]$ except possibly at finitely many points. Then, f is Riemann integrable on $[a, b]$.

The following result will also be needed to prove that d is a metric.

Lemma 1.1 *Let f be a positive continuous function on $[a, b]$. Assume that*

$$\int_a^b f(x)dx = 0.$$

Then, $f(x) = 0$ for all x in $[a, b]$.

Proof of Lemma 1.1
We prove the contrapositive. Assume that f is not identically zero on $[a, b]$. Since f is a positive function there is a c in $[a, b]$ such that $f(c) > 0$. Let $\epsilon = f(c)/2 > 0$. Since f is continuous at c there exists a $\delta_1 > 0$ such that if $|x - c| < \delta_1$, then $|f(x) - f(c)| < \epsilon$. In particular, for x in $(c - \delta_1, c + \delta_1)$,

$$f(x) > f(c) - \epsilon = \frac{f(c)}{2}.$$

Assuming c is strictly between a and b (if $c = a$ or $c = b$ it is easy to modify the argument) we can pick a $\delta_2 > 0$ small enough so that $(c - \delta_2, c + \delta_2) \subset [a, b]$. Let $\delta = \min(\delta_1, \delta_2) > 0$. We have

$$\int_a^b f(x)dx = \int_a^{c-\delta} f(x)dx + \int_{c-\delta}^{c+\delta} f(x)dx + \int_{c+\delta}^b f(x)dx.$$

Since $f \geq 0$ on $[a, b]$ the three integrals on the r.h.s. are positive. Hence,

$$\int_a^b f(x)dx \geq \int_{c-\delta}^{c+\delta} f(x)dx \geq \frac{f(c)}{2}(2\delta) > 0.$$

This proves that $\int_a^b f(x)dx > 0$ and completes the proof of Lemma 1.1.

We are now ready to prove that d is a metric on $C([a, b])$.

Properties (i) and (iii) of Definition 1.1 clearly hold for this metric.

We turn to (ii). Let f and g be continuous and $[a, b]$. Assume that $d(f, g) = 0$. That is,

$$\int_a^b |f(x) - g(x)|dx = 0.$$

Since $|f - g| \geq 0$ and $|f - g|$ is continuous Lemma 1.1 applies. The function $|f - g|$ is identically 0 on $[a, b]$. Therefore, $f = g$ on $[a, b]$. This proves (ii).

We now prove (iv). Take f, g, and h in $C([a, b])$. For every x in $[a, b]$ we have

$$|f(x) - g(x)| = |f(x) - h(x) + h(x) - g(x)| \leq |f(x) - h(x)| + |h(x) - g(x)|.$$

Thus,

$$\int_a^b |f(x) - g(x)|dx| \leq \int_a^b |f(x) - h(x)dx| + \int_a^b |h(x) - g(x)|dx.$$

That is,

$$d(f, g) \leq d(f, h) + d(h, g).$$

This proves (iv) and therefore that d is a metric on $C([a, b])$.

1.4 Euclidean Metric on a Space of Sequences

Consider the space of sequences whose squares are summable. We denote that space by l^2. By definition, $\mathbf{x} = (x_n)$ is an element of l^2 if

$$\sum_{n\geq 1} x_n^2 < +\infty.$$

For instance let $\mathbf{x}$ be defined by $x_n = 1/n$ for every $n \geq 1$, then

$$\sum_{n\geq 1} x_n^2 = \sum_{n\geq 1} \frac{1}{n^2} < +\infty.$$

Hence, $\mathbf{x}$ belongs to l^2.

For $\mathbf{x} = (x_n)$ and $\mathbf{y} = (y_n)$ in l^2 let d be

$$d(\mathbf{x}, \mathbf{y}) = \sqrt{\sum_{n\geq 1} (x_n - y_n)^2}.$$

We will show that d is a metric on l^2. First we check that $d(\mathbf{x}, \mathbf{y})$ is defined for all $\mathbf{x}$ and $\mathbf{y}$ in l^2. This is necessary because d is defined using an infinite series and we need to make sure that the infinite series converges. Given that $\mathbf{x}$ and $\mathbf{y}$ belong to l^2 we need to check that $\mathbf{x} - \mathbf{y}$ also belongs to l^2. We start with the partial sums

$$\sum_{k=1}^{n} (x_k - y_k)^2 = \sum_{k=1}^{n} x_k^2 - 2\sum_{k=1}^{n} x_k y_k + \sum_{k=1}^{n} y_k^2.$$

The equality above shows that if $\sum_{k\geq 1} x_k y_k$ converges so does the series $\sum_{k\geq 1}(x_k - y_k)^2$ (why?). By Cauchy-Schwarz inequality we have

$$\left(\sum_{k=1}^{n} |x_k||y_k|\right)^2 \leq \sum_{k=1}^{n} x_k^2 \sum_{k=1}^{n} y_k^2.$$

Taking square roots yields

$$s_n = \sum_{k=1}^{n} |x_k||y_k| \leq \sqrt{\sum_{k=1}^{n} x_k^2 \sum_{k=1}^{n} y_k^2}.$$

Note now that (s_n) is an increasing sequence and since $\mathbf{x}$ and $\mathbf{y}$ belong to l^2, (s_n) is also bounded above (why?). Hence, (s_n) converges. Therefore, the series $\sum_{k\geq 1} x_k y_k$ converges absolutely. This shows that $d(\mathbf{x}, \mathbf{y})$ is a finite (positive) number for all $\mathbf{x}$ and $\mathbf{y}$ in l^2.

We now check that d is a metric in l^2 using Definition 1.1. Clearly, (i) and (iii) hold. Assume now that

$$d(\mathbf{x}, \mathbf{y}) = \sqrt{\sum_{n\geq 1}(x_n - y_n)^2} = 0.$$

This can happen if and only if $(x_n - y_n)^2 = 0$ for every $n \geq 1$. That is, if and only if $x_n = y_n$ for every $n \geq 1$. Hence, $\mathbf{x} = \mathbf{y}$. This shows (ii).

We now turn to the triangle inequality. Let $\mathbf{x}$, $\mathbf{y}$, and $\mathbf{z}$ be in l^2. By Minkowski Inequality

$$\sqrt{\sum_{k=1}^{n}(x_k - y_k)^2} = \sqrt{\sum_{k=1}^{n}(x_k - z_k + z_k - y_k)^2}$$

$$\leq \sqrt{\sum_{k=1}^{n}(x_k - z_k)^2} + \sqrt{\sum_{k=1}^{n}(z_k - y_k)^2}.$$

We now let n go to infinity on both sides of the inequality to get

$$d(\mathbf{x}, \mathbf{y}) \leq d(\mathbf{x}, \mathbf{z}) + d(\mathbf{z}, \mathbf{y}).$$

This shows the triangle inequality and completes the proof that d is a metric on l^2.

Problems

1. (a) Show that the Cauchy-Schwarz inequality

$$\Big(\sum_{i=1}^{n} a_i b_i\Big)^2 \le \sum_{i=1}^{n} a_i^2 \sum_{i=1}^{n} b_i^2,$$

becomes an equality if and only if there exists a real number t such that for all $i = 1, 2, \ldots, n$ we have

$$a_i + t b_i = 0.$$

What does this tell us about the vectors $\mathbf{a} = (a_1, \ldots, a_n)$ and $\mathbf{b} = (b_1, \ldots, b_n)$?

(b) Show that the Minkowski inequality is an equality if and only if the Cauchy-Schwarz is an equality.

2. Let $\mathbf{a} = (a_1, \ldots, a_n)$ and $\mathbf{b} = (b_1, \ldots, b_n)$ be in $\mathbb{R}^n$. Let

$$d(\mathbf{a}, \mathbf{b}) = \sum_{i=1}^{n} |a_i - b_i|.$$

(a) Show that d is a metric on $\mathbb{R}^n$.
(b) Sketch the graph of the unit ball

$$B = \{\mathbf{a} \in \mathbb{R}^2 : d(\mathbf{0}, \mathbf{a}) \le 1\},$$

where $\mathbf{0} = (0, 0)$.

3. Let $\mathbf{a} = (a_1, \ldots, a_n)$ and $\mathbf{b} = (b_1, \ldots, b_n)$ be in $\mathbb{R}^n$. Let

$$d(\mathbf{a}, \mathbf{b}) = \max_{i=1\ldots n} |a_i - b_i|.$$

(a) Show that d is a metric on $\mathbb{R}^n$.
(b) Sketch the graph of the unit ball

$$B = \{\mathbf{a} \in \mathbb{R}^2 : d(\mathbf{0}, \mathbf{a}) \le 1\},$$

where $\mathbf{0} = (0, 0)$.

4. Let (X, d) be a metric space. Show that for x, y, and z in X we have

$$|d(x, y) - d(x, z)| \le d(z, y).$$

5. (a) Give an example of a positive function f on $[0, 1]$ such that $\int_0^1 f(x)dx = 0$ but f is not identically 0.
(b) Using the example in (a) explain why the Riemann integral metric is not a metric on the space of Riemann integrable functions.

6. Define d on a set X by $d(x, y) = 1$ if $x \neq y$ and $d(x, x) = 0$. This is the so-called *discrete* metric. Show that d is indeed a metric on X.

7. Show that if the series $\sum_{n\geq 1} a_n^2$ and $\sum_{n\geq 1} b_n^2$ converge so does the series $\sum_{n\geq 1} a_n b_n$.

8. Consider the space of sequences l^1 whose absolute values are summable. That is, $\mathbf{x} = (x_n)$ is an element of l^1 if

$$\sum_{n\geq 1} |x_n| < +\infty.$$

For $\mathbf{x} = (x_n)$ and $\mathbf{y} = (y_n)$ in l^1 let d be

$$d(\mathbf{x}, \mathbf{y}) = \sum_{n\geq 1} |x_n - y_n|.$$

(a) Show that $d(\mathbf{x}, \mathbf{y})$ is a finite number for any $\mathbf{x}$ and $\mathbf{y}$ in l^1.
(b) Show that d is a metric on l^1.

2 Sequences in Metric Spaces

2.1 Convergent Sequences

A sequence (a_n) in the metric space (X, d) is a function from $\mathbb{N}$ to X.
We start with the definition of convergence.

Definition 2.1 *Let (X, d) be a metric space. The sequence (a_n) is said to converge to ℓ if for every $\epsilon > 0$ there exists a natural N such that if $n \geq N$, then $d(a_n, \ell) < \epsilon$.*

- Observe that if $X = \mathbb{R}$ and $d(x, y) = |x - y|$, then this definition is exactly the definition we have been using on $\mathbb{R}$.
- On a metric space the limit is unique. That is, if (a_n) converges to ℓ_1 and converges to ℓ_2, then $\ell_1 = \ell_2$. See the problems.

Example 2.1 Let $X = (0, 1]$ with the Euclidean metric. Let $a_n = \frac{1}{n}$ for $n \geq 1$. Note that a_n belongs to X for all $n \geq 1$. However, the sequence (a_n) converges to 0 which is not in X. Hence, the sequence (a_n) does not converge in X. If we take instead $X = [0, 1]$, then the sequence converges in X.

Proposition 2.1 *The sequence (a_n) converges to ℓ in (X, d) if and only if the sequence $(d(a_n, \ell))$ converges to 0 in $(\mathbb{R}, |.|)$.*

Proof of Proposition 2.1
The sequence $(d(a_n, \ell))$ is a sequence of real numbers. By definition it converges to 0 in $(\mathbb{R}, |.|)$ if for every $\epsilon > 0$ there is a natural $N \geq 1$ such that if $n \geq N$ we have

$$|d(a_n, \ell) - 0| < \epsilon.$$

Since $d(a_n, \ell) \geq 0$, we have for $n \geq N$

$$d(a_n, \ell) < \epsilon.$$

This is exactly the definition of (a_n) converging to ℓ in (X, d). Hence, (a_n) converges to ℓ in (X, d) if and only if the sequence $(d(a_n, \ell))$ converges to 0 in $(\mathbb{R}, |.|)$. This completes the proof of Proposition 2.1.

We will now turn to an important property using subsequences. We first recall a definition.

Let (x_n) be a sequence in the metric space (X, d). Let $n_1 < n_2 < \ldots$ be a strictly increasing sequence of natural numbers, then (x_{n_k}) is said to be a **subsequence** of (x_n).

Proposition 2.2 *A sequence (x_n) converges to ℓ if and only if every subsequence of (x_n) converges to ℓ.*

The proof of Proposition 2.2 is an easy generalization of the proof we wrote for the metric space $(\mathbb{R}, |.|)$ (see Theorem 4.1 in Chapter 2) and is left as an exercise.

2.2 *Cauchy Sequences*

The concept of Cauchy sequence is easily generalized to a metric space.

Let (X, d) be a metric space. The sequence (a_n) is said to be **Cauchy** if for every $\epsilon > 0$ there exists a natural N such that if $n \geq N$ and $p \geq N$, then

$$d(a_n, a_p) < \epsilon.$$

Proposition 2.3 *A convergent sequence in (X, d) is Cauchy in (X, d).*

Proof of Proposition 2.3
Let (a_n) be a sequence converging to ℓ in (X, d). Let $\epsilon > 0$. By definition there is a natural $N \geq 1$ such that if $n \geq N$, then

$$d(a_n, \ell) < \frac{\epsilon}{2}.$$

We also have that if $p \geq N$, then

$$d(a_p, \ell) < \frac{\epsilon}{2}.$$

By the triangle inequality

$$d(a_n, a_p) \leq d(a_n, \ell) + d(\ell, a_p) < \frac{\epsilon}{2} + \frac{\epsilon}{2} = \epsilon,$$

for all $n \geq N$ and $p \geq N$. That is, (a_n) is a Cauchy sequence. This completes the proof of Proposition 2.3.

We now show that the converse of Proposition 2.3 does not hold.

Example 2.2 Not every Cauchy sequence is convergent!
Going back to Example 2.1 let $a_n = \frac{1}{n}$ for $n \geq 1$. This sequence converges to 0 in $(\mathbb{R}, |.|)$. By Proposition 2.2 this is a Cauchy sequence in $(\mathbb{R}, |.|)$. It is also a Cauchy sequence in $((0, 1], |.|)$ (why?). However, as pointed out in Example 2.1 it does not converge in $((0, 1], |.|)$.

- Example 2.2 illustrates the fact that a Cauchy sequence that does not converge in some metric space will converge in an expanded metric space.

Recall that a subset A of real numbers is bounded in $(\mathbb{R}, |.|)$ if there exists a k such that $|x| < k$ for all x in A. We now generalize this definition.

The sequence (a_n) is **bounded** in the metric space (X, d) if there exists x in X and $r > 0$ such that for all $n \geq 1$,

$$d(x, a_n) < r.$$

Proposition 2.4 *A Cauchy sequence is bounded.*

Proof of Proposition 2.4
Let (a_n) be a Cauchy sequence in (X, d). There is an $N \geq 1$ such that if $n \geq N$ and $p \geq N$, then

$$d(a_n, a_p) < 1.$$

Let $x = a_N$. We see that for all $n \geq N$, $d(x, a_n) < 1$. On the other hand, the set of real numbers $\{d(x, a_n); n < N\}$ is finite and hence has a maximum m. Let $r = \max(m + 1, 1)$. Then, for all $n \geq 1$

$$d(x, a_n) < r.$$

This completes the proof of Proposition 2.4.

Problems

1. Show that the sequence $a_n = \frac{1}{n}$ for $n \geq 1$ is Cauchy in $((0, 1], |.|)$.

2. Let (a_n) be a real sequence. Consider the two statements.

S1: There exists a real number m such that $|a_n| < m$ for all $n \geq 1$.
S2: There exist real numbers x and $r > 0$ such that $|x - a_n| < r$ for all $n \geq 1$.

Show that S1 and S2 are equivalent.

3. Show that

$$d(x, y) = |\arctan x - \arctan y|$$

is a metric on $\mathbb{R}$.

4. Let $a_n = n$ for $n \geq 1$.

(a) Show that (a_n) is not a Cauchy sequence in $(\mathbb{R}, |.|)$.
(b) Show that (a_n) is a Cauchy sequence in $(\mathbb{R}, d)$ where $d(x, y) = |\arctan x - \arctan y|$.
(c) Show that (a_n) is bounded in $(\mathbb{R}, d)$.

5. In this problem we show that a convergent sequence in a metric space has a unique limit. Assume that (a_n) converges to ℓ_1 and to ℓ_2 in the metric space (X, d).

(a) Show that for every $\epsilon > 0$, $d(\ell_1, \ell_2) < \epsilon$.
(b) Show that $\ell_1 = \ell_2$.

6. Show that a convergent sequence in a metric space is bounded.

7. Assume that (a_n) is a Cauchy sequence in a metric space and that it has a convergent subsequence. Show that (a_n) converges.

8. Prove Proposition 2.2.

9. Let (x_n) and (y_n) be two sequences in a metric space (X, d). Assume that (x_n) converges to some ℓ. Assume also that $(d(x_n, y_n))$ converges to 0. Show that (y_n) converges to ℓ as well.

3 Complete Metric Spaces

We have seen above that a Cauchy sequence in a metric space (X, d) need not converge. A metric space (X, d) is said to be **complete** if every Cauchy sequence in (X, d) converges in (X, d).

3.1 Complete and Incomplete Euclidean Spaces

Theorem 3.1 *The metric space* $(\mathbb{R}, |.|)$ *is complete.*

Theorem 6.1 in Chapter 2 shows that every Cauchy sequence in $(\mathbb{R}, |.|)$ converges. This proves Theorem 3.1.

Example 3.1 Recall that $\mathbb{Q}$ is the set of rational numbers. The metric space $(\mathbb{Q}, |.|)$ is not complete. That is, there are sequences in $\mathbb{Q}$ that are Cauchy but do not converge in $\mathbb{Q}$. See, for instance Example 6.1 in Chapter 2.

Theorem 3.2 *For every* $n \geq 1$ *the metric space* $\mathbb{R}^n$ *equipped with the Euclidean metric is complete.*

Proof of Theorem 3.2
The case $n = 1$ is Theorem 3.1. The proof for $n \geq 2$ is actually a consequence of Theorem 3.1. We will prove Theorem 3.2 for $n = 2$ and leave the general case (which is an easy generalization) to the Problems.

Let $(\mathbf{a}_n)$ be a Cauchy sequence in $(\mathbb{R}^2, d)$ where d is the Euclidean metric. Let $\mathbf{a}_n = (x_n, y_n)$. We will show first that (x_n) and (y_n) are Cauchy sequences in $(\mathbb{R}, |.|)$. Observe that for $n \geq 1$ and $p \geq 1$ we have

$$(x_n - x_p)^2 \leq (x_n - x_p)^2 + (y_n - y_p)^2$$

and therefore

$$\sqrt{(x_n - x_p)^2} \leq \sqrt{(x_n - x_p)^2 + (y_n - y_p)^2}.$$

Hence,

$$|x_n - x_p| \leq d(\mathbf{a}_n, \mathbf{a}_p).$$

Similarly,

$$|y_n - y_p| \leq d(\mathbf{a}_n, \mathbf{a}_p).$$

Since $(\mathbf{a}_n)$ is Cauchy for $\epsilon > 0$ there is a natural N such that if $n \geq N$ and $p \geq N$, then

$$d(\mathbf{a}_n, \mathbf{a}_p) < \epsilon.$$

Therefore for $n \geq N$ and $p \geq N$ we have

$$|x_n - x_p| < \epsilon \text{ and } |y_n - y_p| < \epsilon.$$

That is, the sequences (x_n) and (y_n) are Cauchy in $(\mathbb{R}, |.|)$. By Theorem 3.1 the sequences (x_n) and (y_n) converge in $(\mathbb{R}, |.|)$ to some x and y, respectively. Let $\mathbf{a} = (x, y)$. We have

$$d(\mathbf{a}_n, \mathbf{a}) = \sqrt{(x_n - x)^2 + (y_n - y)^2}.$$

Since $(x_n - x)^2$ and $(y_n - y)^2$ converge to 0 (why?) so does $\sqrt{(x_n - x)^2 + (y_n - y)^2}$ (why?) and hence $d(\mathbf{a}_n, \mathbf{a})$ converges to 0. We have proved that a Cauchy sequence must converge in $(\mathbb{R}^2, d)$. The proof of Theorem 3.2 is complete for $n = 2$.

As the next example shows not all metrics make the set $\mathbb{R}$ complete!

Example 3.2 Consider $\mathbb{R}$ with the arctan metric. That is, for x and y in $\mathbb{R}$ define

$$d(x, y) = |\arctan(x) - \arctan(y)|.$$

It is easy to check that d is a metric on $\mathbb{R}$. For $n \geq 1$, let $a_n = n$. Recall that

$$\lim_{x\to\infty} \arctan(x) = \frac{\pi}{2},$$

in $(\mathbb{R}, |.|)$. Hence, for $\epsilon > 0$ there exists N such that if $n \geq N$, then

$$|\arctan(n) - \frac{\pi}{2}| < \frac{\epsilon}{2}.$$

Therefore, for $n \geq N$ and $p \geq N$

$$\begin{aligned} d(a_n, a_p) &= |\arctan(n) - \arctan(p)| \\ &\leq |\arctan(n) - \frac{\pi}{2}| + |\frac{\pi}{2} - \arctan(p)| \\ &< \frac{\epsilon}{2} + \frac{\epsilon}{2} = \epsilon. \end{aligned}$$

That is, (a_n) is Cauchy in $(\mathbb{R}, d)$.

We will now prove that (a_n) does not converge. By contradiction assume that (a_n) converges to ℓ in $(\mathbb{R}, d)$. Then,

$$d(a_n, \ell) = |\arctan(n) - \arctan(\ell)|$$

converges to 0 in $(\mathbb{R}, |.|)$. On the other hand, $d(a_n, \ell)$ converges to $|\frac{\pi}{2} - \arctan(\ell)|$. By the uniqueness of the limit we must have

$$|\frac{\pi}{2} - \arctan(\ell)| = 0.$$

That is, $\arctan(\ell) = \frac{\pi}{2}$. There is no such ℓ in $\mathbb{R}$. Hence, we found a Cauchy sequence in $(\mathbb{R}, d)$ which does not converge. This is not a complete space.

3.2 Complete and Incomplete Metrics on the Space of Continuous Functions

Consider $C([a, b])$ the set of real-valued continuous functions on $[a, b]$. Let

$$d_u(f, g) = \max_{x \in [a,b]} |f(x) - g(x)|.$$

In Chapter 5 we defined uniform convergence as follows. The sequence (f_n) converges uniformly to f on $[a, b]$ if for every $\epsilon > 0$ there exists a natural N such that for all x in $[a, b]$

$$|f_n(x) - f(x)| < \epsilon.$$

We showed that (f_n) converges uniformly to f if and only if (m_n) converges to 0 where

$$m_n = \sup_{x \in [a,b]} |f_n(x) - f(x)|.$$

In the particular case where f_n and f are continuous the Extreme Value Theorem applies on the closed bounded interval $[a, b]$ and we have

$$m_n = \sup_{x \in [a,b]} |f_n(x) - f(x)| = \max_{x \in [a,b]} |f_n(x) - f(x)| = d_u(f_n, f).$$

That is, uniform convergence in $C([a, b])$ is the same as convergence on the metric space $(C([a, b]), d_u)$. By the same token, a sequence of continuous functions is uniformly Cauchy on $[a, b]$ if and only if it is a Cauchy sequence in $(C([a, b]), d_u)$.

Theorem 3.3 *The metric space* $(C([a, b]), d_u)$ *is complete.*

Corollary 2.1 in Chapter 5 shows that a sequence of continuous functions which is uniformly Cauchy on $[a, b]$ converges uniformly to a continuous function on $[a, b]$. Hence, Corollary 2.1 implies Theorem 3.3.

We now show that $C([a, b])$ is not complete under the Riemann integral metric.

Consider $C([a, b])$ with the metric

$$d_r(f, g) = \int_a^b |f(x) - g(x)| dx.$$

Theorem 3.4 *The metric space $(C([a, b]), d_r)$ is not complete.*

Proof of Theorem 3.4
We will give an example of a Cauchy sequence which does not converge in this metric space. We will show the theorem for a particular a and b. The general case follows from the particular case. Let $a = 0$ and $b = 2$. Define the sequence (f_n) as follows. For $n \geq 1$ let

$$f_n(x) = x^n \text{ for } x \in [0, 1] \text{ and } f_n(x) = 1 \text{ for } x \in (1, 2].$$

For every n, the function f_n is continuous on $[0, 2]$ (why?). By definition,

$$d_r(f_n, f_p) = \int_0^2 |f_n(x) - f_p(x)|dx = \int_0^1 |x^n - x^p|dx.$$

Hence,

$$d_r(f_n, f_p) \leq \int_0^1 (x^n + x^p)dx = \frac{1}{n+1} + \frac{1}{p+1}.$$

By the Archimedean property there exists a natural N such that $\frac{1}{N+1} < \frac{\epsilon}{2}$. Therefore, for $n \geq N$ and $p \geq N$,

$$d_r(f_n, f_p) \leq \frac{1}{n+1} + \frac{1}{p+1} < \frac{\epsilon}{2} + \frac{\epsilon}{2} = \epsilon.$$

Hence, (f_n) is Cauchy in $(C([0, 2]), d_r)$.

We now show that (f_n) does not converge in $(C([0, 2]), d_r)$. By contradiction assume that (f_n) does converge to some f in this metric space. We need f to be continuous on $[0, 2]$ and $d(f_n, f)$ to converge to 0. We have

$$d_r(f_n, f) = \int_0^2 |f_n(x) - f(x)|dx = \int_0^1 |x^n - f(x)|dx + \int_1^2 |1 - f(x)|dx.$$

Since both terms are positive in the r.h.s. we need both terms to converge to 0. Note that $\int_1^2 |1 - f(x)|dx$ is a constant so it has to be equal to 0. Since $|1 - f|$ is continuous on $[1, 2]$ (why?) and its integral is 0 the function $|1 - f|$ is identically 0 on $[1, 2]$, see Lemma 1.1. Hence,

$$f(x) = 1 \text{ for all } x \in [1, 2].$$

We now turn to $\int_0^1 |x^n - f(x)|dx$. By the Triangle inequality, for all real numbers a and b,

$$|a| - |b| \leq |a - b| \leq |a| + |b|.$$

Therefore, for all x in $[0, 1]$

$$|f(x)| - |x^n| \leq |f(x) - x^n| \leq |f(x)| + |x^n|.$$

We now take integrals across inequalities to get

$$\int_0^1 (|f(x)| - |x^n|)dx \leq \int_0^1 (|f(x) - x^n|)dx \leq \int_0^1 (|f(x)| + |x^n|)dx.$$

Thus,

$$\int_0^1 |f(x)|dx - \frac{1}{n+1} \leq \int_0^1 (|f(x) - x^n|)dx \leq \int_0^1 |f(x)|dx + \frac{1}{n+1}.$$

By the squeezing principle we get that

$$\lim_{n\to\infty} \int_0^1 (|f(x) - x^n|)dx = \int_0^1 |f(x)|dx.$$

As noted above this limit is 0. Hence, $\int_0^1 |f(x)|dx = 0$. We use again Lemma 1.1 to show that

$$f(x) = 0 \text{ for all } x \in [0, 1].$$

Since $f(x) = 1$ on $[1, 2]$ the function f is not continuous at $x = 1$. We have a contradiction. The sequence (f_n) does not converge in $(C([0, 2]), d_r)$. This metric space is not complete. The proof of Theorem 3.4 is complete.

- It is always possible to embed a metric space (X, d) which is not complete into a bigger space which is. The smallest complete space that contains (X, d) is called the completion of (X, d) (see Kreiszig (1978) for instance). For example, the metric space $(\mathbb{R}, |.|)$ is the completion of $(\mathbb{Q}, |.|)$.

Problems

1. Prove Theorem 3.2 for any $n \geq 3$. (Follow the proof given for $n = 2$).

2. Use Cauchy-Schwarz inequality to show for real numbers $a_1, a_2, \ldots, a_n$

$$(\sum_{i=1}^{n} a_i)^2 \leq n \sum_{i=1}^{n} a_i^2.$$

3. Consider the following three metrics on $(\mathbb{R}^n, d)$. Let $\mathbf{a} = (a_1, \ldots, a_n)$ and $\mathbf{b} = (b_1, \ldots, b_n)$ in $\mathbb{R}^n$. Define

$$d_1(\mathbf{a}, \mathbf{b}) = \max_{1 \le i \le n} |a_i - b_i|.$$

$$d_2(\mathbf{a}, \mathbf{b}) = \sum_{i=1}^{n} |a_i - b_i|.$$

$$d_3(\mathbf{a}, \mathbf{b}) = \sqrt{\sum_{i=1}^{n} (a_i - b_i)^2}.$$

For all $\mathbf{a}$ and $\mathbf{b}$ in $\mathbb{R}^n$ prove the following inequalities.

(a) Show that $d_1(\mathbf{a}, \mathbf{b}) \le d_2(\mathbf{a}, \mathbf{b})$.
(b) Show that $d_2(\mathbf{a}, \mathbf{b}) \le \sqrt{n} d_3(\mathbf{a}, \mathbf{b})$. (Use Problem 2.).
(c) Show that $d_3(\mathbf{a}, \mathbf{b}) \le \sqrt{n} d_1(\mathbf{a}, \mathbf{b})$.

4. Two metrics d_1 and d_2 on some set X are said to be **equivalent** if there exist $\alpha > 0$ and $\beta > 0$ such that for all a and b in X

$$\alpha d_1(a, b) < d_2(a, b) < \beta d_1(a, b).$$

Use the inequalities of Problem 3. to show that the three metrics defined there are equivalent on $\mathbb{R}^n$.

5. Let d_1 and d_2 be two equivalent metrics on X (see problem 4.). Show that (X, d_1) is complete if and only if (X, d_2) is complete.

6. Show that $(\mathbb{R}^n, d_1)$ and $(\mathbb{R}^n, d_2)$ are complete where d_1 and d_2 are defined in Problem 3..

7. The discrete metric on a set X is defined by $d(x, y) = 1$ if $x \neq y$ and $d(x, x) = 0$.

(a) Show that if (a_n) is a Cauchy sequence with respect to the discrete metric, then there exists a natural N and a point c in X such that if $n \ge N$, then $a_n = c$.
(b) Show that X with the discrete metric is complete.

8. For $n \ge 1$ let $I_n = [a_n, b_n]$ where (a_n) and (b_n) are two sequences in $\mathbb{R}$. Assume also that for all $n \ge 1$ we have $I_{n+1} \subset I_n$. Let

$$I = \bigcap_{n \ge 1} I_n.$$

In this problem we show that I is not empty.

(a) Show that the sequence (a_n) is increasing and bounded above.
(b) Show that (a_n) converges to some ℓ in $(\mathbb{R}, |.|)$.
(c) Show that ℓ belongs to I. (Show that $a_m \leq a_{m+n} \leq b_{m+n} \leq b_m$ for all natural numbers n and m.)
(d) Give an example showing that I may be empty if we do not assume that the I_n are closed.
(e) Give an example showing that I may be empty if we do not assume that the I_n are bounded.

9. Let X be the set of bounded real sequences. That is, the sequence $\mathbf{s} = (s_n)$ belongs to X if and only if there exists a k such that for all $n \geq 1$ we have $|s_n| < k$. Let $\mathbf{a} = (a_n)$ and $\mathbf{b} = (b_n)$ be in X and define

$$d(\mathbf{a}, \mathbf{b}) = \sup_{n \geq 1} |a_n - b_n|.$$

(a) Show that $d(\mathbf{a}, \mathbf{b})$ is a finite number for all $\mathbf{a}$ and $\mathbf{b}$ in X.
(b) Show that d is a metric on X.

10. In this problem we show that the metric space (X, d) defined in problem 9. is complete.

Let $(\mathbf{s}_n)$ be a sequence in X. For fixed n, $\mathbf{s}_n$ is a sequence of real numbers that we will denote by $\mathbf{s}_n = (s_n(i)) = (s_n(1), s_n(2), \ldots)$.

Assume that $(\mathbf{s}_n)$ is Cauchy in (X, d).

(a) Show that for $\epsilon > 0$ there exists N such that if $n \geq N$ and $p \geq N$, then for all $i \geq 1$ we have

$$|s_n(i) - s_p(i)| < \epsilon.$$

(b) Show that for fixed $i \geq 1$ the sequence of real numbers $(s_n(i))$ is Cauchy in $(\mathbb{R}, |.|)$.
(c) Show that for fixed $i \geq 1$ the sequence $(s_n(i))$ converges to some $s(i)$ in $(\mathbb{R}, |.|)$. This defines a $\mathbf{s} = (s(i))$.
(d) Show that $\mathbf{s}$ is in X (i.e., it is a bounded sequence).
(e) Show that $(\mathbf{s}_n)$ converges to $\mathbf{s}$ in (X, d). This completes the proof that (X, d) is complete.

11. Consider the space of sequences l^1 whose absolute values are summable. That is, $\mathbf{x} = (x_n)$ is an element of l^1 if

$$\sum_{n \geq 1} |x_n| < +\infty.$$

For $\mathbf{x} = (x_n)$ and $\mathbf{y} = (y_n)$ in l^1 let d be

$$d(\mathbf{x}, \mathbf{y}) = \sum_{n \geq 1} |x_n - y_n|.$$

We already proved in a problem above that d is a metric on l^1. Show that (l^1, d) is complete.

12. Consider the space of sequences l^2 whose squares are summable. We defined a metric d on this space, see Section 1.4. Show that (l^2, d) is complete.

13. Consider the space of continuous functions on $[0, 1]$, $C([0, 1]$, equipped with the uniform metric d. Show that the set of polynomials is dense in $C([0, 1])$. That is, show that for any f in $C([0, 1]$ and any $r > 0$ there exists a polynomial p such that

$$d(f, p) < r.$$

(Use the Bernstein polynomials of Chapter 5.)

14. Let X be the set of all polynomials on $[0, 1]$ and let d be defined by

$$d(f, g) = \max_{x \in [0,1]} |f(x) - g(x)|.$$

(a) Let

$$f_n(x) = \sum_{k=0}^{n} \frac{x^k}{k!}.$$

Show that (f_n) converges uniformly on $[0, 1]$.

(b) Show that (X, d) is not a complete space.

Chapter 8
Topology in a Metric Space

In this chapter we define several topological concepts such as open, closed, and compact sets. They will turn out to be quite important when studying functions on a metric space.

1 Open and Closed Sets

Let (X, d) be a metric space. We define the **open ball** centered at $a \in X$ with radius $r > 0$ by

$$B(a, r) = \{x \in X : d(a, x) < r\}.$$

Definition 1.1 *A set $A \subset X$ is said to be open in the metric space (X, d) if for every $a \in A$ there exists $r > 0$ such that $B(a, r) \subset A$.*

In words, every point of an open set is the center of an open ball which is entirely in the set. Intuitively, an open set has points only in its interior. There can be no point on the boundary.

- The empty set $\emptyset$ and the whole space X are open sets in (X, d) for any metric d.

Proposition 1.1 *An open ball $B(a, r)$ is an open set.*

- In several proofs below we will need to show that $A \subset B$. The typical inclusion proof consists in showing that if x belongs to A, then it belongs to B. To show that a set A is the same as a set B we will show that $A \subset B$ and $B \subset A$.

Proof of Proposition 1.1
Let b in $B(a, r)$. We need to find $r_1 > 0$ such that $B(b, r_1) \subset B(a, r)$. Let

$$r_1 = r - d(a, b).$$

R. B. Schinazi, *From Classical to Modern Analysis*,
https://doi.org/10.1007/978-3-319-94583-5_8

We have that $r_1 > 0$ (why?). Let x be in $B(b, r_1)$. Then,

$$d(b, x) < r_1 = r - d(a, b),$$

That is, $d(a, b) + d(b, x) < r$. By the triangle inequality we get

$$d(a, x) \leq d(a, b) + d(b, x) < r.$$

Hence, x belongs to $B(a, r)$. This shows that if x be in $B(b, r_1)$, then x belongs to $B(a, r)$. That is, $B(b, r_1) \subset B(a, r)$. This completes the proof of Proposition 1.1.

Example 1.1 Consider the Euclidean metric space $(\mathbb{R}, |.|)$. In this particular case

$$B(a, r) = \{x \in \mathbb{R} : |a - x| < r\} = (a - r, a + r).$$

That is, the open ball $B(a, r)$ is the open interval $(a - r, a + r)$. Moreover, by Proposition 1.1, $(a - r, a + r)$ is an open set in $(\mathbb{R}, |.|)$.

Recall that the open intervals in $\mathbb{R}$ are the sets $(-\infty, +\infty)$, $(-\infty, a)$, (a, b), and $(b, +\infty)$ for real numbers a and b. We next show that open intervals are indeed open with respect to the Euclidean metric on $\mathbb{R}$.

Proposition 1.2 *Open intervals are open sets in* $(\mathbb{R}, |.|)$.

Proof of Proposition 1.2
We will prove that (a, b) is open for any $a < b$ and leave the other cases to the Problems. Let

$$x = \frac{a + b}{2} \text{ and } r = \frac{b - a}{2}.$$

The open ball $B(x, r)$ is then the set $\{y : |y - x| < r\}$. This is exactly the set (a, b). That is, (a, b) is an open ball and by Proposition 1.1 is an open set. This completes the proof of Proposition 1.2.

Example 1.2 Consider the interval $(0, 1]$ in $(\mathbb{R}, |.|)$. This is not an open set. For 1 belongs to $(0, 1]$ but it is not possible to find r such that

$$B(1, r) = (1 - r, 1 + r) \subset (0, 1].$$

Hence, $(0, 1]$ is not open.

We now give a way to construct open sets in $\mathbb{R}^n$ using open sets in $(\mathbb{R}, |.|)$.

Proposition 1.3 *Let $n \geq 2$ be a natural number. Let $U_1, U_2, \ldots, U_n$ be open sets in $(\mathbb{R}, |.|)$. Then, the Cartesian product $U_1 \times U_2 \times \cdots \times U_n$ is open in $\mathbb{R}^n$ equipped with the Euclidean metric.*

Proof of Proposition 1.3

Consider $\mathbf{a} = (a_1, a_2, \ldots, a_n)$ in $U = U_1 \times U_2 \times \cdots \times U_n$. We want to find $r > 0$ such that the open ball (with respect to the Euclidean metric) $B(\mathbf{a}, r)$ is included in U. Let i be a natural between 1 and n. Observe that a_i belongs to U_i which is open in $(\mathbb{R}, |.|)$. Hence, there exists an open ball $B(a_i, r_i)$ entirely in U_i for some $r_i > 0$. This is true for every i between 1 and n. Define

$$r = \min_{1 \le i \le n} r_i.$$

We have that $r > 0$ (why?). We will now show that $B(\mathbf{a}, r) \subset U$. Let $\mathbf{y} = (y_1, \ldots, y_n)$ be in $B(\mathbf{a}, r)$. Then,

$$d(\mathbf{a}, \mathbf{y}) = \sqrt{\sum_{i=1}^{n} (a_i - y_i)^2} < r.$$

Since for every i, $|a_i - y_i| \le d(\mathbf{a}, \mathbf{y})$ (why?) we have

$$|a_i - y_i| < r \le r_i.$$

That is, y_i belongs to $B(a_i, r_i)$ which is a subset of U_i. Hence, y_i belongs to U_i. Since this is true for every i, $\mathbf{y}$ belongs to U. This proves that $B(\mathbf{a}, r) \subset U$. The proof of Proposition 1.3 is complete.

The following properties give a way to construct more open sets.

Proposition 1.4

(i) *The union of a finite or infinite collection of open sets is open.*
(ii) *The intersection of a finite collection of open sets is open.*

Note that (i) holds for an infinite collection but (ii) does not hold for an infinite collection of open sets. See the problems.

Proof of Proposition 1.4
We prove (i) first. Let $A_1, A_2 \ldots$ be a finite or infinite collection of open sets. Let x be in

$$A = \bigcup_{j \ge 1} A_j.$$

Then, x belongs to A_j for some j. Since A_j is open there exists $r_x > 0$ such that $B(x, r_x) \subset A_j$. Therefore, $B(x, r_x) \subset A$. Note that the same argument applies for finite or infinite unions of open sets. This completes the proof of (i).

We now prove (ii). Let $A_1, A_2 \dots, A_n$ be a finite collection of open sets. Let

$$B = \bigcap_{j=1}^{n} A_j.$$

Assume a is in B. Then, for every $j = 1, \dots, n, a \in A_j$. Since every A_j is open we can find $r_j > 0$ such that $B(a, r_j) \subset A_j$ for $j = 1, \dots, n$. Let $r = \min_{1 \le j \le n} r_j$. Note that $r > 0$. Observe that for $j = 1, \dots, n$, $B(a, r) \subset B(a, r_j)$. Hence, $B(a, r) \subset A_j$ for $j = 1, \dots, n$ and therefore $B(a, r) \subset B$. This completes the proof of Proposition 1.4.

The following is an important characterization of open sets.

Corollary 1.1 *A set is open if and only if it is a union of open balls.*

The proof of Corollary 1.1 is an easy application of Proposition 1.4 and is left to the reader.

We now turn to the notion of closed set. The **complement** of a set A is defined by

$$A^c = \{x \in X : x \notin A\}.$$

That is, A^c is the set of all the points that are not in A.

Definition 1.2 *A set A is said to be* ***closed*** *in (X, d) if A^c is open in (X, d).*

- Note that a set need not be open or closed. It can very well be neither.
- The sets X and $\emptyset$ are closed and open.

Example 1.3 The set $[0, 1)$ is neither open nor closed in $(\mathbb{R}, |.|)$.

Let $A = [0, 1)$. The complement of A is $A^c = (-\infty, 0) \cup [1, +\infty)$. Hence, 1 belongs to A^c but it is not possible to find an open ball $B(1, r)$ inside A^c. Hence, A^c is not open and therefore A is not closed. Since A is not open either (why?) it is neither open nor closed.

Example 1.4 Closed intervals $(-\infty, a]$, $[a, b]$ and $[b, +\infty)$ are closed sets in $(\mathbb{R}, |.|)$.

Consider $(-\infty, a]$, its complement is $(a, +\infty)$. Since this is an open set in $(\mathbb{R}, |.|)$, $(-\infty, a]$ is closed. The other cases are left to the problems.

The following set properties will be helpful.

Proposition 1.5 (de Morgan's Laws) *Let $A_1, A_2, \dots,$ be subsets of X. Then,*

$$\Big(\bigcup_{i \ge 1} A_i\Big)^c = \bigcap_{i \ge 1} A_i^c,$$

$$\Big(\bigcap_{i \ge 1} A_i\Big)^c = \bigcup_{i \ge 1} A_i^c.$$

Proof of Proposition 1.5
We prove the first law. The element x belongs to $\left(\bigcup_{i\geq 1} A_i\right)^c$ if and only it does not belong to $\bigcup_{i\geq 1} A_i$. That is, if and only it belongs to none of the A_i. This is the same as saying that x belongs to all the A_i^c. This is equivalent to x belonging to $\bigcap_{i\geq 1} A_i^c$. This completes the proof of the first law. The second law is left as an exercise.

Proposition 1.6

(i) *The union of a finite collection of closed sets is closed.*
(ii) *The intersection of a finite or infinite collection of closed sets is closed.*

The proof of Proposition 1.6 is an immediate consequence of Proposition 1.4 and de Morgan's laws. The proof is left as an exercise.

The following characterization of closed sets using sequences is quite useful.

Proposition 1.7 *A set A is closed in a metric space (X, d) if and only if every sequence in A that converges has its limit in A.*

- Not every sequence in a closed set converges! However, if it converges, then the limit must belong to the set. The converse of this property is also true.

Proof of Proposition 1.7
By contradiction assume that A is closed and that there exists a sequence (a_n) in A that converges to ℓ but ℓ does not belong to A. Using that A^c is open and ℓ belongs to A^c there exists $r > 0$ such that the open ball $B(\ell, r) \subset A^c$. Since the sequence (a_n) converges to ℓ there exists n such that if $n \geq N$, then $d(a_n, \ell) < r$. Therefore, for $n \geq N$, $a_n \in B(\ell, r)$. But $B(\ell, r) \subset A^c$. Hence, $a_n \in A^c$ for all $n \geq N$. However, we have assumed that the sequence (a_n) is in A. We have a contradiction. Thus, ℓ must belong to A. This proves the direct implication.

We now turn to the converse. By contradiction assume that A is not closed. Then A^c is not open. Since A^c is not open there exists an ℓ in A^c such that for every $r > 0$ the open ball $B(\ell, r)$ is not included in A^c. So for every $r > 0$ there exists a point in $B(\ell, r)$ which is not in A^c and therefore is in A. Since the last statement is true for any $r > 0$ we can pick $r = \frac{1}{n}$ where n is a natural number. Hence, for every $n \geq 1$ there exists a point a_n in $B(\ell, \frac{1}{n}) \cap A$. The sequence (a_n) is therefore in A and for every $n \geq 1$

$$d(a_n, \ell) < \frac{1}{n}.$$

That is, the sequence (a_n) converges to ℓ. We have found a sequence (a_n) in A that converges outside of A. We have a contradiction. The set A must be closed. This completes the proof of Proposition 1.7.

The following property will be used many times.

Proposition 1.8 *Consider a metric space (X, d). Let (x_n) be a sequence in X converging to some ℓ. Let a be a point in X. Then the sequence $(d(a, x_n))$ converges in $(\mathbb{R}, |.|)$ to $d(a, \ell)$.*

Proof of Proposition 1.8
By the triangle inequality,

$$|d(a, x_n) - d(a, \ell)| \leq d(x_n, \ell).$$

Since (x_n) converges to ℓ, the sequence of real numbers $(d(x_n, \ell))$ converges to 0. Hence, $(d(a, x_n) - d(a, \ell))$ converges to 0 (why?). Therefore, $(d(a, x_n))$ converges to $d(a, \ell)$. This completes the proof of Proposition 1.8.

Example 1.5 Consider a metric space (X, d). Let a in X and $r > 0$. Define the **closed ball** $\tilde{B}(a, r)$ by

$$\tilde{B}(a, r) = \{x \in X : d(a, x) \leq r\}.$$

Then, the set $\tilde{B}(a, r)$ is closed in (X, d). We now prove this fact.

Consider a sequence (x_n) in $\tilde{B}(a, r)$ that converges to some ℓ. By Proposition 1.8, the numerical sequence $(d(a, x_n))$ converges to $d(a, \ell)$. Since $d(a, x_n) \leq r$ for every $n \geq 1$, letting n go to infinity we get

$$d(a, \ell) \leq r.$$

That is, ℓ belongs to $\tilde{B}(a, r)$. Hence, a sequence in $\tilde{B}(a, r)$ that converges must converge inside $\tilde{B}(a, r)$. By Proposition 1.7, this proves that $\tilde{B}(a, r)$ is closed in (X, d).

Example 1.6 Show that the circle

$$C = \{(x, y) \in \mathbb{R}^2 : x^2 + y^2 = 1\}$$

is a closed set of $\mathbb{R}^2$ equipped with the Euclidean metric.

Let d be the Euclidean metric in $\mathbb{R}^2$. Then, the set C can also be written as

$$C = \{\mathbf{a} \in \mathbb{R}^2 : d(\mathbf{0}, \mathbf{a}) = 1\}$$

where $\mathbf{0} = (0, 0)$. Let $(\mathbf{a_n})$ be a sequence in C that converges to some $\mathbf{a}$. Since $d(\mathbf{0}, \mathbf{a_n}) = 1$ for every $n \geq 1$, letting n go infinity and using Proposition 1.8 we get

$$d(\mathbf{0}, \mathbf{a}) = 1.$$

Hence, $\mathbf{a}$ belongs to C. This proves that C is closed in $(\mathbb{R}^2, d)$.

Let A be a set in the metric space (X, d). Next we define the **closure** of A which is denoted by $\bar{A}$.

Definition 1.3 *Let A be a set in the metric space (X, d). We define the closure of A as the set consisting of all the limits of sequences in A. That is, x belongs to $\bar{A}$ if and only if there exists a sequence (a_n) in A converging to x.*

- Let a be in A. Define the sequence $a_n = a$ for all $n \geq 1$. Then, (a_n) converges to a. Hence, a belongs to $\bar{A}$. This proves that A is always a subset of $\bar{A}$.
- Using Proposition 1.7 it is easy to see that $\bar{A} = A$ if and only if A is closed.

Consider the set $A = [0, 1)$ in $(\mathbb{R}, |.|)$. Then, 1 belongs to $\bar{A}$ (why?). In fact, it is easy to check that $\bar{A} = [0, 1]$.

Next we give an important characterization of the closure of a set.

Proposition 1.9 *Let A be a set in the metric space (X, d). The closure of A consists of all the points x in X such that for every $r > 0$, $A \cap B(x, r) \neq \emptyset$.*

In words, x belongs to $\bar{A}$ if and only if every open ball centered at x intersects A.

Proof of Proposition 1.9
Assume that x belongs to $\bar{A}$. There exists a sequence (a_n) in A converging to x. Thus, for every $r > 0$ there exists N such that if $n \geq N$, then $d(x, a_n) < r$. That is, a_n belongs to $B(x, r)$ for all $n \geq N$. Hence, $A \cap B(x, r) \neq \emptyset$.

We now turn to the converse. Let x be such that $A \cap B(x, r) \neq \emptyset$ for every $r > 0$. Hence, for any $n \geq 1$, we can find an a_n in $B(x, \frac{1}{n}) \cap A$ (why?). Therefore, (a_n) is a sequence in A and for any $n \geq 1$,

$$d(x, a_n) < \frac{1}{n}.$$

Thus, the sequence (a_n) converges to x. This proves the converse.

Proposition 1.10 *The closure $\bar{A}$ of a set A is the smallest closed set containing A. That is, if $A \subset F$ and F is closed, then $\bar{A} \subset F$.*

Proof of Proposition 1.10
We first prove that $\bar{A}$ is closed. Take b not in $\bar{A}$. Then, by Proposition 1.9 there exists $r > 0$ such that $B(b, r) \cap A = \emptyset$. We are now going to show that the bigger set $B(b, r) \cap \bar{A}$ is also empty. By contradiction, assume there is an x in $B(b, r) \cap \bar{A}$. Then, x belongs to $B(b, r)$ and there exists a sequence (a_n) in A that converges to x. Hence, the sequence $(d(b, a_n))$ converges to $d(b, x) < r$. Since $B(b, r) \cap A = \emptyset$,

$$d(b, a_n) \geq r,$$

for every $n \geq 1$. Letting n go to infinity in the last inequality we get $d(b, x) \geq r$. This is a contradiction. Hence, $B(b, r) \cap \bar{A}$ is empty. That is, $(\bar{A})^c$ is open and therefore $\bar{A}$ is closed.

We now show that $\bar{A}$ is the smallest closed set containing A. Take F to be closed and containing A. We will show that F contains $\bar{A}$ as well. Let x be in $\bar{A}$. Then, for every $r > 0$, $B(x, r) \cap A \neq \emptyset$. Hence, $B(x, r) \cap F \neq \emptyset$. Therefore, x belongs to $\bar{F}$.

Since F is closed $\bar{F} = F$. Therefore, x belongs to F. This shows that $\bar{A} \subset F$ and completes the proof of Proposition 1.10.

Problems

1. Show that if $0 < r < s$, then

$$B(a, r) \subset B(a, s).$$

2. Show that if $A \subset B$, then $B^c \subset A^c$.

3. Show that a union of open intervals is an open set in $(\mathbb{R}, |.|)$.

4.

(a) Prove that $(b, +\infty)$ is open in $(\mathbb{R}, |.|)$.
(b) Prove that $(-\infty, a)$ is open in $(\mathbb{R}, |.|)$.

5. Consider $\mathbb{R}^2$ equipped with the Euclidean metric.

(a) Show that $(a, b) \times (c, d)$ is an open set.
(b) Give an example of an open set which is not a Cartesian product of open intervals.

6. In this problem we show that an infinite intersection of open sets is not necessarily open.

Consider the intervals

$$F_n = (-\frac{1}{n}, \frac{1}{n}).$$

(a) Show that for $n \geq 1$ the set F_n is open in $(\mathbb{R}, |.|)$.
(b) Show that

$$\bigcap_{n \geq 1} F_n = \{0\}.$$

(c) Is $\bigcap_{n \geq 1} F_n$ an open set?

7. Show that

$$\bigcup_{n \geq 1} (0, 1 - \frac{1}{n}) = (0, 1).$$

8. In the proof of (ii) of Proposition 1.4 we show that a finite intersection of open sets is open. Where does the argument break down for an infinite intersection of open sets?

9. Recall that two metrics d_1 and d_2 on some set X are said to be equivalent if there exist $\alpha > 0$ and $\beta > 0$ such that for all a and b in X

$$\alpha d_1(a, b) < d_2(a, b) < \beta d_1(a, b).$$

Show that a set A is open in (X, d_1) if and only if it is open in (X, d_2).

10. Consider the set $C([0, 1])$ of continuous functions on $[0, 1]$.

(a) Describe the open ball $B(0, 1)$ (where 0 is the function identically 0) in $C([0, 1])$ equipped with the uniform metric (i.e., $d(f, g) = \max_{0 \leq x \leq 1} |f(x) - g(x)|$).
(b) What is the open ball $B(0, 1)$ with respect to the Riemann integral metric (i.e., $d(f, g) = \int_0^1 |f(x) - g(x)| dx$)?

11. Show that $[a, b]$ is a closed set in $(\mathbb{R}, |.|)$.

12. Show that the set

$$A = \{(x, y) \in \mathbb{R}^2 : x^2 + y^2 > 1\}$$

is an open set of $\mathbb{R}^2$ equipped with the Euclidean metric.

13. Show that the set

$$B = \{(x, y) \in \mathbb{R}^2 : x^2 - y^2 = 1\}$$

is a closed set of $\mathbb{R}^2$ equipped with the Euclidean metric.

14.

(a) Show that the closure of $[0, 1)$ in $(\mathbb{R}, |.|)$ is $[0, 1]$.
(b) Show that the closure of the open ball $B(\mathbf{0}, 1)$ in $\mathbb{R}^2$ equipped with Euclidean metric is the closed ball $\tilde{B}(\mathbf{0}, 1)$.

15. Show that $A = \bar{A}$ if and only if A is closed.

16. Interior of a set. Consider a set A in a metric space (X, d). Let the interior of A be denoted by $\mathring{A}$ and be defined by

$$\mathring{A} = \{x \in A : B(x, r) \subset A \text{ for some } r > 0\}.$$

In words, x is in the interior of A if there is on open ball centered at x which is entirely in A.

(a) Show that the interior of $[0, 1)$ in $(\mathbb{R}, |.|)$ is $(0, 1)$.
(b) Show that the interior of $\{1\}$ in $(\mathbb{R}, |.|)$ is $\emptyset$.
(c) Show that the interior of the closed ball $\tilde{B}(\mathbf{0}, 1)$ in $\mathbb{R}^2$ equipped with Euclidean metric is the open ball $B(\mathbf{0}, 1)$.

17. See the definition of the interior $\mathring{A}$ of a set A in Problem 16..

(a) Show that A is open in a metric space (X, d) if and only if $A = \mathring{A}$.
(b) Show that if U is open in (X, d) and $U \subset A$, then $U \subset \mathring{A}$.
(c) Use (b) to show that $\mathring{A}$ is open in (X, d).
(d) Show that $\mathring{A}$ is the largest open set contained in A.

18. Prove Corollary 1.1.

2 Compactness

2.1 General Properties

We start with the definition of a compact set.

Definition 2.1 *Let K be a set in the metric space (X, d). The set K is compact if every sequence in K has a convergent subsequence whose limit is an element of K.*

The following result gives an important example of compactness.

Proposition 2.1 *For any $a < b$, the closed bounded interval $[a, b]$ is compact in the metric space $(\mathbb{R}, |.|)$.*

Proof of Proposition 2.1
Let (x_n) be a sequence in $[a, b]$. By Bolzano-Weierstrass Theorem (x_n) has a subsequence (x_{n_k}) that converges to some ℓ in $(\mathbb{R}, |.|)$. Moreover, for every $k \geq 1$

$$a \leq x_{n_k} \leq b.$$

Letting k go to infinity in the two inequalities above yield $a \leq \ell \leq b$. Hence, (x_{n_k}) has a limit ℓ which is in $[a, b]$. Thus, $[a, b]$ is compact in $(\mathbb{R}, |.|)$. This completes the proof of Proposition 2.1.

Example 2.1 The set $\mathbb{R}$ is not compact in $(\mathbb{R}, |.|)$.

To show this it is enough to find a sequence (x_n) in $\mathbb{R}$ which has no convergent subsequence. Let $x_n = n$ for every $n \geq 1$. No subsequence of (x_n) is bounded (why?) and therefore no subsequence converges.

We now turn to two important properties of compact sets.

Proposition 2.2 *If the set K is compact in the metric space (X, d), then it is closed in (X, d).*

Proof of Proposition 2.2
Assume that K is compact. Let (x_n) be a sequence in K that converges. It has a convergent subsequence (x_{n_k}) that converges to some $\ell \in K$. Since (x_{n_k}) is a subsequence of the convergent sequence (x_n), the whole sequence (x_n) converges to ℓ as well (why?). Hence, a convergent sequence in K must converge to a limit in K. This proves that K is closed and completes the proof of Proposition 2.2.

Before stating the second property of compact sets we need a definition.

Definition 2.2 *A set A is said to be bounded in the metric space (X, d) if there exist x in X and $r > 0$ in $\mathbb{R}$ such that A is a subset of the open ball $B(x, r)$.*

Proposition 2.3 *If the set K is compact in the metric space (X, d), then it is bounded in (X, d).*

Proof of Proposition 2.3
We do a proof by contradiction. Assume that K is not bounded in (X, d). Then for any x in X and any $n \geq 1$ there is a_n in K such that

$$d(x, a_n) \geq n.$$

Since K is compact there exists a subsequence (a_{n_k}) that converge to some ℓ in K. Hence, the sequence $\big(d(x, a_{n_k})\big)$ converges to $d(x, \ell)$ (why?). We also have for $k \geq 1$

$$d(x, a_{n_k}) \geq n_k.$$

Hence, the sequence $(d(x, a_{n_k}))$ is not bounded and cannot converge. We have a contradiction. Therefore, K is bounded. This completes the proof of Proposition 2.3.

Putting together the two last propositions we get.

Corollary 2.1 *If the set K is compact in the metric space (X, d), then it is closed and bounded in (X, d).*

In general the converse of Corollary 2.1 is not true. We will give a counter-example at the end of this section. It is however true in the following important particular case.

2.2 Compact Sets in the Euclidean Spaces

The following result characterizes all the compact sets in $\mathbb{R}^n$.

Theorem 2.1 *Let $n \geq 1$, a set is compact in $\mathbb{R}^n$ equipped with the Euclidean metric if and only if it is closed and bounded.*

We already know that a compact is closed and bounded. We need to show that in the Euclidean space $\mathbb{R}^n$ a set which is closed and bounded is necessarily compact. We will need two intermediate results in order to prove this.

Lemma 2.1 *Let K be a compact set in a metric space (X, d). Consider a subset A of K. If A is closed, then A is compact in (X, d).*

Proof of Lemma 2.1
Let (x_n) be a sequence in A. Since (x_n) is also in K which is compact there exists a subsequence (x_{n_k}) that converges to some ℓ in K. But (x_{n_k}) is in A and A is closed. Recall that a convergent sequence in a closed set must converge in the set. Hence, ℓ belongs to A. This proves that A is compact and completes the proof of Lemma 2.1.

Lemma 2.2 *Let $n \geq 1$ and $a_1 < b_1, a_2 < b_2, \ldots, a_n < b_n$. The Cartesian product $[a_1, b_1] \times [a_2, b_2] \times \cdots \times [a_n, b_n]$ is compact in $\mathbb{R}^n$ equipped with the Euclidean metric.*

Proof of Lemma 2.2
We do an induction proof on n. For $n = 1$ this has already been proved in Proposition 2.1. Assume that the statement is true for a natural n. That is,

$$K_n = [a_1, b_1] \times [a_2, b_2] \times \cdots \times [a_n, b_n]$$

is compact in $\mathbb{R}^n$. Let

$$K_{n+1} = K_n \times [a_{n+1}, b_{n+1}]$$

where $a_{n+1} < b_{n+1}$. Let $(\mathbf{a_j})$ be a sequence in K_{n+1}. Then for every j, $\mathbf{a}_j = (\mathbf{c}_j, d_j)$ where $\mathbf{c}_j$ has n coordinates and d_j is a real number. Since the sequence $(\mathbf{c}_j)$ is in K_n which is compact there exists a subsequence $(\mathbf{c}_{j_k})$ which converges to some $\mathbf{c}$ in K_n. On the other hand, the subsequence (d_{j_k}) is in the compact set $[a_{n+1}, b_{n+1}]$. Therefore, there is a sub-subsequence of (d_j) that we denote $(d_{j'_k})$ that converges to some d in $[a_{n+1}, b_{n+1}]$. Finally, observe that $(\mathbf{c}_{j'_k})$ is a subsequence of $(\mathbf{c}_{j_k})$ and therefore converges to $\mathbf{c}$. Using the Euclidean metric it is easy to see that $(\mathbf{a}_{j'_k})$ converges to $(\mathbf{c}, d)$ (see the Problems). Hence, K_{n+1} is compact. This concludes the proof of Lemma 2.2.

Proof of Theorem 2.1
Let A be a closed and bounded set in $\mathbb{R}^n$ equipped with the Euclidean metric d. Since A is bounded it must be a subset of an open ball $B(\mathbf{x}, r)$ where $\mathbf{x}$ is in $\mathbb{R}^n$ and $r > 0$. Let $(x_1, \ldots, x_n)$ be the coordinates of $\mathbf{x}$. If $\mathbf{a} = (a_1, \ldots, a_n)$ is in A, then

$$d(\mathbf{x}, \mathbf{a}) = \sqrt{\sum_{i=1}^{n} (x_i - a_i)^2} < r.$$

Hence, for every $i = 1, \ldots, n$, $|x_i - a_i| < r$. Therefore $\mathbf{a}$ belongs to the cartesian product $\Pi_{i=1}^{n}(x_i - r, x_i + r)$. Thus, the set A is included in $\prod_{i=1}^{n}(x_i - r, x_i + r)$ which is itself included in

$$K = \prod_{i=1}^{n}[x_i - r, x_i + r].$$

By Lemma 2.2, K is a compact set. Since A is closed (by hypothesis) and is a subset of the compact set K we can conclude that A is compact as well by Lemma 2.1. This completes the proof of Theorem 2.1.

2.3 A Closed Ball Which Is Not Compact

We give here an example showing that the converse of Corollary 2.1 is not true. We will exhibit a closed and bounded set which is not compact.

Consider the set l^2 of real sequences $\mathbf{x} = (x_n)$ such that

$$\sum_{n \geq 1} x_n^2 < +\infty.$$

Let

$$d(\mathbf{x}, \mathbf{y}) = \sqrt{\sum_{n \geq 1}(x_n - y_n)^2},$$

for $\mathbf{x}$ and $\mathbf{y}$ in l^2. We have seen in Section 1.4, Chapter 7 that d is a metric on l^2. We now define the sequence $(\mathbf{y}_n)$ in l^2 as follows. For every $n \geq 1$, $\mathbf{y}_n$ is itself a sequence of real numbers

$$\mathbf{y}_n = (y_n(1), y_n(2), \ldots) \text{ where } y_n(n) = 1 \text{ and } y_n(k) = 0 \text{ for all } k \neq n.$$

Note that for every $n \geq 1$, $\mathbf{y}_n$ is in l^2. Consider now the closed unit ball in l^2 centered at $\mathbf{0}$ (i.e., the sequence identically 0).

$$\tilde{B}(\mathbf{0}, 1) = \{\mathbf{x} \in l^2 : d(\mathbf{0}, \mathbf{x}) \leq 1\}.$$

Clearly,

$$d(\mathbf{0}, \mathbf{y}_n) = \sqrt{\sum_{k \geq 1}(0 - y_n(k))^2} = 1.$$

Hence, for every $n \geq 1$, $\mathbf{y}_n$ belongs to $\tilde{B}(\mathbf{0}, 1)$. Observe now that for $n \neq p$

$$d(\mathbf{y}_n, \mathbf{y}_p) = \sqrt{\sum_{k \geq 1} (y_n(k) - y_p(k))^2} = \sqrt{2}.$$

Therefore, $(\mathbf{y}_n)$ is not Cauchy. Moreover, no subsequence of $(\mathbf{y}_n)$ can be Cauchy (why?). Hence, no subsequence converges. This shows that $\tilde{B}(\mathbf{0}, 1)$ is not compact.

- We have shown that the closed unit ball is compact in $\mathbb{R}^n$ and not compact in l^2. It turns out that in a normed vector space the closed unit ball is compact if and only if the space is finite dimensional! See section 2.5 in Kreiszig (1978).

2.4 *The Cantor Set*

In this section we describe the Cantor set. This is a set of real numbers with many remarkable properties. It is a good example to keep in mind when checking ideas about real numbers.

We start with

$$S_0 = [0, 1].$$

Remove the middle third interval $(\frac{1}{3}, \frac{2}{3})$ from S_0 to get

$$S_1 = [0, \frac{1}{3}] \cup [\frac{2}{3}, 1].$$

Remove the middle thirds of each of these intervals to get

$$S_2 = [0, \frac{1}{9}] \cup [\frac{2}{9}, \frac{3}{9}] \cup [\frac{6}{9}, \frac{7}{9}] \cup [\frac{8}{9}, 1].$$

Continuing this procedure we get a sequence (S_n) of sets in $(\mathbb{R}, |.|)$. Note that for every $n \geq 0$, S_n is the union of 2^n closed and bounded intervals. Moreover, each interval included in S_n has length 3^{-n}.

We define the Cantor set C by the infinite intersection

$$C = \bigcap_{n \geq 1} S_n.$$

We start by showing that the Cantor set is not empty.

- The Cantor set has infinitely many elements.

  As noted before S_n is the union of 2^n closed and bounded intervals. Consider the $2 \times 2^n = 2^{n+1}$ distinct real numbers that are endpoints of these 2^n intervals. They belong to $\mathcal{C}$ by construction. This is so because at every step we eliminate the middle third intervals. So an endpoint is never eliminated by this process and will end up in $\mathcal{C}$. For instance, $0, \frac{1}{3}, \frac{2}{3}, 1$ all belong to $\mathcal{C}$. Since the number of endpoints grows without bound the Cantor set must be infinite.
- The Cantor set is a compact set of $(\mathbb{R}, |.|)$.

  Each S_n is closed as a finite union of closed sets. Hence, $\mathcal{C}$ is closed as an infinite intersection of closed sets. Since $\mathcal{C}$ is also bounded (why?) it is a compact set of $(\mathbb{R}, |.|)$.

  We already know that $\mathcal{C}$ is infinite. Next we are more precise.
- The Cantor set $\mathcal{C}$ is uncountable.

  An uncountable set is a set that cannot be put in a sequence. If we try to label the elements of $\mathcal{C}$ by $x_1, x_2, \ldots$ we run out of labels! More precisely, if we take any sequence (x_n) in $\mathcal{C}$ we can always find an element of $\mathcal{C}$ which is not in the sequence (x_n).

  Next we give the sketch of the proof that $\mathcal{C}$ is uncountable. Every x in $[0, 1]$ has a ternary expansion

$$x = \sum_{n \geq 1} \frac{a_n}{3^n},$$

where for each $n \geq 1$, a_n belongs to $\{0, 1, 2\}$. It turns out that the cantor set $\mathcal{C}$ is the set of numbers whose ternary expansion can be written without any 1. That is, $x \in \mathcal{C}$ if and only if

$$x = \sum_{n \geq 1} \frac{a_n}{3^n} \text{ where } a_n = 0 \text{ or } a_n = 2 \text{ for every } n \geq 1.$$

We now explain why. Note that if $a_1 = 1$, then x must be in $(\frac{1}{3}, \frac{2}{3})$. This is the interval we remove to obtain S_1 from S_0. If $a_1 \neq 1$ and $a_2 = 1$, then x must be in

$$(\frac{1}{9}, \frac{2}{9}) \text{ if } a_1 = 0$$

or x must be in

$$(\frac{7}{9}, \frac{8}{9}) \text{ if } a_1 = 2.$$

These are precisely the two intervals we remove to obtain S_2 from S_1. More generally, for a given x, if $a_1 \neq 1, a_2 \neq 1, \ldots, a_{k-1} \neq 1$ and a_k is 1, then x must belong to one of the open intervals we remove to obtain S_k from S_{k-1}.

This shows that x is in $\mathcal{C}$ if and only the sequence (a_n) can be chosen so that none of its terms is a 1. Hence, there is a one-to-one correspondence between the Cantor set and the set of sequences with only 0's and 2's. Since this set of sequences is uncountable (see for instance Chapter 9 in Schinazi (2011)) so is the Cantor set.

- No interval is contained in the Cantor set.

The critical observation is that if I is an interval in $\mathbb{R}$ and if $x < y$ belong to I, then the interval $[x, y]$ is a subset of I.

By contradiction assume that there is an interval I containing $x < y$ in the Cantor set. Then, $[x, y] \subset I$. Hence, for every $n \geq 0$, $[x, y] \subset S_n$. Since S_n is a union of 2^n disjoint intervals and $[x, y]$ is an interval, $[x, y]$ need to be included in one (and only one) of these 2^n intervals (why?). However, each one of these intervals has length 3^{-n}. Hence,

$$0 \leq y - x \leq 3^{-n}.$$

Since this is true for all $n \geq 0$ we can let n go to infinity to get that $x = y$. We have a contradiction. No interval with more than one element is included in $\mathcal{C}$.

Problems

1. Show that a singleton is always a compact set.

2. Show that $(0, 1]$ is not compact in $(\mathbb{R}, |.|)$.

3. Show that if $(\mathbf{a}_j)$ converges to $\mathbf{a}$ in $\mathbb{R}^n$ and $(\mathbf{b}_j)$ converges to $\mathbf{b}$ in $\mathbb{R}^p$, then $((\mathbf{a}_j, \mathbf{b}_j))$ converges to $(\mathbf{a}, \mathbf{b})$ in $\mathbb{R}^{n+p}$, where all the metrics are Euclidean.

4. Assume that K_1 and K_2 are compact sets in $(\mathbb{R}, |.|)$. Show that the Cartesian product $K_1 \times K_2$ is compact in $\mathbb{R}^2$ equipped with the Euclidean metric.

5. Consider a sequence of compact sets $K_1, K_2 \ldots, K_n$ in a metric space (X, d). Let

$$K = \bigcup_{i=1}^{n} K_i.$$

Let (x_n) be a sequence in K.

(a) Show that there must be infinitely many terms of (x_n) in at least one of the K_i's.
(b) Use (a) to show that K is compact.
(c) Give an example that shows that an infinite union of compact sets need not be compact.

6. Consider an infinite sequence of nonempty compact sets $K_1, K_2 \ldots$, in $(\mathbb{R}, |.|)$. Assume also that for all $n \geq 1$, $K_{n+1} \subset K_n$. Let

$$K = \bigcap_{n \geq 1} K_n.$$

(a) Show that K is compact in $(\mathbb{R}, |.|)$.

Since for every n we assume that K_n is not empty we can pick an element in every K_n that we denote by x_n. This defines a sequence (x_n).

(b) Show that (x_n) has a subsequence (x_{n_j}) that converges to some ℓ.
(c) Let $m \geq 1$. Show that for all $j \geq m$, x_{n_j} belongs to K_m.
(d) Show that ℓ belongs to K. This proves that K is not empty.

7. In this problem we show that the Cantor set $\mathcal{C}$ has no **isolated points**. Mathematically, this means that if x belongs to $\mathcal{C}$, we can find a sequence (x_n) in $\mathcal{C}$ with the following properties. For every n, $x_n \neq x$ and (x_n) converges to x.

(a) Let x be in $\mathcal{C}$. Show that for every $n \geq 0$, x belongs to an interval I_n whose length is 3^{-n}. Let x_n be an endpoint of I_n which is not x.
(b) Show that (x_n) belongs to $\mathcal{C}$.
(c) Show that for every $n \geq 0$, $x_n \neq x$ and (x_n) converges to x.

8. Assume that A is a bounded subset of $\mathbb{R}^n$ equipped with the Euclidean metric. Show that the closure $\bar{A}$ of A is compact.

9. Show that a compact metric space is complete. That is, take (x_n) to be a Cauchy sequence in (K, d) where K is compact and show that (x_n) converges to some ℓ in K.

10. Let K be a compact set in the metric space (X, d). Let a be in K^c. Consider the set

$$A = \{d(x, a); x \in K\}.$$

(a) Show that A has a greatest lower bound m.
(b) Show that there exists b in K so that $m = d(b, a)$. That is, the minimum distance between a and K is attained.
(c) Show that if K is not compact the property need not be true.

11. Let K be a compact set in the metric space (X, d) and B be a closed set. Consider the set

$$A = \{d(x, y); x \in K, y \in B\}.$$

(a) Show that A has a greatest lower bound m.
(b) Prove that $m > 0$ if and only if $K \cap B = \emptyset$.

Chapter 9
Continuity on Metric Spaces

1 Definition and Examples

Definition 1.1 *Let (X, d_1) and (Y, d_2) be two metric spaces. Let $f : X \longrightarrow Y$ and a an element of X. The function f is said to be continuous at a if for every sequence (x_n) converging to a the sequence $(f(x_n))$ converges to $f(a)$.*

Note that (x_n) and $(f(x_n))$ converge with respect to the metrics d_1 and d_2, respectively. So the continuity of f is with respect to these metrics. The function need not be continuous with respect to other metrics.

Example 1.1 Let $X = \mathbb{R}^3$, $Y = \mathbb{R}$ and let d_1 and d_2 be the Euclidean metrics on $\mathbb{R}^3$ and $\mathbb{R}$, respectively. Let P be the polynomial

$$P(x, y, z) = xyz - x^2 + 5y - 1.$$

Let $\mathbf{a} = (a_1, a_2, a_3)$ be a point of $\mathbb{R}^3$. Let $\mathbf{x}_n = (c_n, d_n, e_n)$ be a sequence in $\mathbb{R}^3$ converging to $\mathbf{a}$. We have that

$$d_2(a_1, c_n) = |a_1 - c_n| \le \sqrt{(a_1 - c_n)^2 + (a_2 - d_n)^2 + (a_3 - e_n)^2} = d_1(\mathbf{a}, \mathbf{x}_n).$$

Since $(\mathbf{x}_n)$ converges to $\mathbf{a}$ we have that (c_n) converges to a_1. Similarly, (d_n) converges to a_2 and (e_n) converges to a_3. Using operations on limits for sequences in $(\mathbb{R}, |.|)$ we get,

$$P(\mathbf{x}_n) = c_n d_n e_n - c_n^2 + 5d_n - 1 \text{ converges to } a_1 a_2 a_3 - a_1^2 + 5a_2 - 1 = P(\mathbf{a}).$$

This shows that P is continuous everywhere on $\mathbb{R}^3$.

- The method used in Example 1.1 can be used to show that any polynomial on $\mathbb{R}^n$ is continuous everywhere with respect to the Euclidean metrics.

R. B. Schinazi, *From Classical to Modern Analysis*,
https://doi.org/10.1007/978-3-319-94583-5_9

As we see next, operations on limits on $\mathbb{R}$ can be used to show continuity for a wide class of numerical functions.

Proposition 1.1 *Let f and g be defined on X with values in $\mathbb{R}$. Assume that X is equipped with a metric d and $\mathbb{R}$ with the Euclidean metric. Assume that f and g are continuous at a. Then, $f+g$, $f \times g$ are continuous at a. Moreover, if $g(a) \neq 0$, then $\frac{f}{g}$ is continuous at a.*

Proof of Proposition 1.1
The proofs follow immediately from the operations on limits in $(\mathbb{R}, |.|)$. Assume that (a_n) is a sequence converging to a in (X, d); then by definition of continuity we have that $(f(a_n))$ and $(g(a_n))$ converge to $f(a)$ and $g(a)$, respectively. By operations on limits we have that

$$(f(a_n) + g(a_n)) \text{ converges to } f(a) + g(a),$$

and

$$(f(a_n)g(a_n)) \text{ converges to } f(a)g(a).$$

Hence, $f+g$ and $f \times g$ are continuous at a.

Assuming that $g(a) \neq 0$ we must have $g(a_n) \neq 0$ for $n \geq N$ for some N (why?). By operations on limits we have

$$(\frac{f(a_n)}{g(a_n)}) \text{ converges to } \frac{f(a)}{g(a)}.$$

This completes the proof of Proposition 1.1.

- Proposition 1.1 can be used to show that any rational function (i.e., the ratio of two polynomials) on $\mathbb{R}^n$ is continuous with respect to the Euclidean metrics wherever the function is defined.

Proposition 1.2 *Let f and g be defined on possibly different metric spaces. Assume that the range of f is included in the domain of g. Assume also that g is continuous at a and f is continuous at $g(a)$. Then $g \circ f$ is continuous at a.*

Proof of Proposition 1.2
Assume that (a_n) is a sequence converging to a. By continuity of f at a, the sequence $(f(a_n))$ converges to $f(a)$. By continuity of g at $f(a)$ we have that $\big(g(f(a_n))\big)$ converges to $g(f(a))$. Hence, $g \circ f$ is continuous at a. This completes the proof of Proposition 1.2.

Example 1.2 Consider the function $h : \mathbb{R}^2 \longrightarrow \mathbb{R}$ such that $h(x, y) = \sqrt{x^2 + y^2}$. Show that h is continuous everywhere.

Note that $h = g \circ f$ where $f(x, y) = x^2 + y^2$ and $g(z) = \sqrt{z}$. The range of f is $[0, +\infty)$ which is the domain of g. Since f is a polynomial it is continuous at any $\mathbf{a}$

in $\mathbb{R}^2$. Since $f(\mathbf{a}) \geq 0$ and g is continuous everywhere on $[0, +\infty)$, g is continuous at $f(\mathbf{a})$. Hence, by Proposition 1.2 $h = g \circ f$ is continuous at any $\mathbf{a}$ in $\mathbb{R}^2$.

Example 1.3 Let (X, d) be a metric space. Let a be in X. Define the function f as

$$f(x) = d(a, x).$$

Show that f is a continuous function on X.

Note that f has real values. We show continuity with respect to d and the Euclidean metric. Let (b_n) be a sequence in X converging to some b. By the triangle inequality,

$$|f(b_n) - f(b)| = |d(a, b_n) - d(a, b)| \leq d(b, b_n).$$

Hence, $(f(b_n))$ converges to $f(b)$ (why?). Therefore, f is continuous at b.

We now turn to an equivalent definition of continuity.

Proposition 1.3 *Let (X, d_1) and (Y, d_2) be two metric spaces. Let $f : X \longrightarrow Y$ and a an element of X. The function f is continuous at a if and only if for every $\epsilon > 0$ there is a $\delta > 0$ such that if $d_1(a, x) < \delta$, then $d_2(f(a), f(x)) < \epsilon$.*

Proof of Proposition 1.3
For the direct implication we prove the contrapositive. Assume that there is $\epsilon > 0$ such that for all $\delta > 0$ there is an x such that $d_1(a, x) < \delta$ and $d_2(f(a), f(x)) \geq \epsilon$. We use the preceding statement in the following way. For $n \geq 1$, let $\delta = \frac{1}{n}$, then there is x_n such that

$$d_1(a, x_n) < \frac{1}{n} \text{ and } d_2(f(a), f(x_n)) \geq \epsilon.$$

Therefore, there exists a sequence (x_n) that converges to a (why?) and such that $(f(x_n))$ does not converge to $f(a)$ (why not?). Thus, f is not continuous at a.

We now prove the converse. Let $\epsilon > 0$. There exists a $\delta > 0$ such that if $d_1(a, x) < \delta$, then $d_2(f(a), f(x)) < \epsilon$. Let (x_n) be a sequence converging to a. Since (x_n) converges to a there exists a natural N such that if $n \geq N$, then $d_1(a, x_n) < \delta$. Hence, $d_2(f(a), f(x_n)) < \epsilon$ for $n \geq N$. Therefore, $(f(x_n))$ converges to $f(a)$. This proves that f is continuous at a. This completes the proof of Proposition 1.3.

2 Image and Inverse Image

We introduce some new notation. Let f be a function from X to Y. Let A be a subset of X. The **image** of A under f is the set

$$f(A) = \{f(x) : x \in A\}.$$

Note that $f(A)$ is a subset of Y. Note also that $f(X)$ is the range of f.

Let B be a subset of Y. The **inverse image** of B under f is the set

$$f^{-1}(B) = \{x \in X : f(x) \in B\}.$$

Note that $f^{-1}(B)$ is a subset of X. The notation f^{-1} is also used to denote the inverse function of f (if it exists!). However, the definition $f^{-1}(B)$ does not assume that f has an inverse. There is no inconsistency between the two notations. If f does have an inverse, then it is easy to check that for any b in Y, $f^{-1}(b)$ is the same using either definition.

Example 2.1 Let $X = Y = \mathbb{R}$ and $f(x) = x^2$. Let $A = [-1, 1]$, then

$$f([-1, 1]) = \{x^2 : x \in [-1, 1]\}.$$

If x is in $[-1, 1]$, then $0 \le x^2 \le 1$. Conversely, if $0 \le x^2 \le 1$, then taking square roots yields $|x| \le 1$. Hence, x is in $[-1, 1]$. In summary,

$$f([-1, 1]) = [0, 1].$$

Example 2.2 Let $X = Y = \mathbb{R}$ and $f(x) = x^2$. Let $B = \{4\}$. Then

$$f^{-1}(B) = \{x : x^2 = 4\}.$$

Hence,

$$f^{-1}(B) = \{-2, 2\}.$$

Proposition 2.1 *Let $f : X \longrightarrow Y$ and let $B_1, B_2, \ldots$ be a sequence of subsets of Y. We have*

$$f^{-1}\Big(\bigcup_{i\ge1} B_i\Big) = \bigcup_{i\ge1} f^{-1}(B_i),$$

$$f^{-1}\Big(\bigcap_{i\ge1} B_i\Big) = \bigcap_{i\ge1} f^{-1}(B_i),$$

$$\big(f^{-1}(B_1)\big)^c = f^{-1}(B_1^c).$$

In words, the inverse image commutes with unions, intersections, and complements. On the other hand, the direct image commutes only with unions. See the Problems.

Proof of Proposition 2.1
We prove only the first statement. The other two statements are left to the Problems. Let x be in $f^{-1}\big(\bigcup_{i\geq 1} B_i\big)$. This means that $f(x)$ is in $\bigcup_{i\geq 1} B_i$. This is equivalent to $f(x)$ being in B_j for some $j \geq 1$. This means that x is in $f^{-1}(B_j)$ for some $j \geq 1$. This is the same as x being in $\bigcup_{i\geq 1} f^{-1}(B_i)$. Hence,

$$f^{-1}\Big(\bigcup_{i\geq 1} B_i\Big) = \bigcup_{i\geq 1} f^{-1}(B_i).$$

3 Continuity and Open Sets

The following is a very useful characterization of continuity.

Theorem 3.1 *Let $f : X \longrightarrow Y$ where (X, d_1) and (Y, d_2) are metric spaces. The function f is continuous everywhere on X if and only if $f^{-1}(V)$ is open in (X, d_1) for every open set V in (Y, d_2).*

Proof of Theorem 3.1
First the direct implication. Assume that f is continuous on X and let V be an open set in Y. Let a be in $f^{-1}(V)$. Then $f(a)$ belongs to V which is open. Hence, there exists an $\epsilon > 0$ such that the open ball $B(f(a), \epsilon) \subset V$. Since f is continuous at a, by Proposition 1.3, there exists a $\delta > 0$ such that if $d_1(a, x) < \delta$, then $d_2(f(a), f(x)) < \epsilon$. The last statement can be written as if $x \in B(a, \delta)$, then $f(x) \in B(f(a), \epsilon)$. Since $f(x) \in B(f(a), \epsilon)$ is the same as $x \in f^{-1}(B(f(a), \epsilon))$, this shows the following inclusion

$$B(a, \delta) \subset f^{-1}\left(B(f(a), \epsilon)\right).$$

Since $B(f(a), \epsilon) \subset V$ we have

$$B(a, \delta) \subset f^{-1}(B(f(a), \epsilon)) \subset f^{-1}(V).$$

This shows that for any a in $f^{-1}(V)$ we can find an open ball $B(a, \delta) \subset f^{-1}(V)$. Hence, $f^{-1}(V)$ is an open set. This proves the direct implication.

We now prove the converse. Let a be in X. Let $\epsilon > 0$ and consider $B(f(a), \epsilon)$ which is an open ball in Y. Note that a belongs to $f^{-1}(B(f(a), \epsilon))$ (why?) and by assumption the set $f^{-1}(B(f(a), \epsilon))$ is open in X. Hence, there exists an open ball centered at a entirely inside $f^{-1}(B(f(a), \epsilon))$. That is, there is a $\delta > 0$ such that

$$B(a, \delta) \subset f^{-1}(B(f(a), \epsilon)).$$

Hence, if $x \in B(a, \delta)$, then $x \in f^{-1}(B(f(a), \epsilon))$. Therefore, if $d_1(a, x) < \delta$, then $f(x) \in B(f(a), \epsilon)$. That is, $d_2(f(a), f(x)) < \epsilon$. By Proposition 1.3, f is continuous at a. This completes the proof of Theorem 3.1.

Example 3.1 Let $a \neq 0$ and $b \neq 0$ be two real numbers. Show that the set

$$A = \{(x, y) : \frac{x^2}{a^2} + \frac{y^2}{b^2} < 1\}$$

is an open set in $\mathbb{R}^2$ equipped with the Euclidean metric.

Let $f(x, y) = \frac{x^2}{a^2} + \frac{y^2}{b^2}$. Then A can be written as

$$A = f^{-1}((-\infty, 1)).$$

Since the set $(-\infty, 1)$ is open in $(\mathbb{R}, |.|)$ and f is a continuous function (why?) A is open in $\mathbb{R}^2$ equipped with the Euclidean metric.

An easy consequence of Theorem 3.1 is the following.

Corollary 3.1 *Let $f : X \longrightarrow Y$ where (X, d_1) and (Y, d_2) are metric spaces. The function f is continuous everywhere on X if and only if $f^{-1}(C)$ is closed in (X, d_1) for every closed set C in (Y, d_2).*

The proof of Corollary 3.1 is left to the problems.

The next example gives a new proof that an open ball is open.

Example 3.2 Show that an open ball $B(a, r)$ in a metric space (X, d) is an open set.

Let $f(x) = d(a, x)$. We have

$$B(a, r) = \{x \in X : d(a, x) < r\} = f^{-1}((-\infty, r)).$$

Since the set $(-\infty, r)$ is open in $(\mathbb{R}, |.|)$ and f is continuous by Example 1.3 the set $f^{-1}((-\infty, r))$ is open in (X, d).

Example 3.3 Let $a \neq 0$ and $b \neq 0$ be two real numbers. Show that the set

$$B = \{(x, y) : \frac{x^2}{a^2} + \frac{y^2}{b^2} \leq 1\}$$

is a closed set in $\mathbb{R}^2$ equipped with the Euclidean metric.

Let $f(x, y) = \frac{x^2}{a^2} + \frac{y^2}{b^2}$. Then A can be written as

$$B = f^{-1}((-\infty, 1]).$$

Since the set $(-\infty, 1]$ is closed in $(\mathbb{R}, |.|)$ and f is a continuous function B is closed in $\mathbb{R}^2$ equipped with the Euclidean metric.

4 Continuity and Compactness

A continuous inverse image of an open set is open. For compact sets it is instead the direct image that plays that role.

Theorem 4.1 *Let $f : X \longrightarrow Y$ where (X, d_1) and (Y, d_2) are metric spaces. Let f be continuous on a compact set K in X. Then, $f(K)$ is compact in Y.*

Proof of Theorem 4.1
Let (b_n) be a sequence in $f(K)$. We need to show the existence of a subsequence of (b_n) that converges to a limit in $f(K)$. Since b_n belongs to $f(K)$, $b_n = f(a_n)$ for some a_n in K. This defines a sequence (a_n) in K. Since K is compact there is a subsequence (a_{n_k}) that converges to some ℓ in K. By the continuity of f on K,

$$(f(a_{n_k})) \text{ converges to } f(\ell).$$

Note that $f(a_{n_k}) = b_{n_k}$ for all $k \geq 1$ and that $f(\ell)$ belongs to $f(K)$ (why?). Hence, there is a subsequence of (b_n) that converges to a limit in $f(K)$. This completes the proof of Theorem 4.1.

Definition 4.1 *Let $f : X \longrightarrow Y$ where (X, d_1) and (Y, d_2) are metric spaces. The function f is said to be* ***uniformly continuous*** *on X if for every $\epsilon > 0$ there exists a $\delta > 0$ such that if*

$$d_1(x, y) < \delta \textit{ then } d_2(f(x), f(y)) < \epsilon.$$

It is clear that uniform continuity implies continuity. The converse is not true. Note first that f is not uniformly continuous if and only if there exists an $\epsilon > 0$ such that for any $\delta > 0$ there are x and y such that $d_1(x, y) < \delta$ and $d_2(f(x), f(y)) \geq \epsilon$. By taking $\delta = \frac{1}{n}$ we get the existence of two sequences (x_n) and (y_n) such that for every $n \geq 1$,

$$d_1(x_n, y_n) < \frac{1}{n} \text{ and } d_2(f(x_n), f(y_n)) \geq \epsilon.$$

We see that f is NOT uniformly continuous on X if and only if we can find two sequences (x_n) and (y_n) in X such that the sequence $(d_1(x_n, y_n))$ converges to 0 and the sequence $(d_2(f(x_n), f(y_n)))$ is bounded away from 0.

Example 4.1 Consider $f(x) = 1/x$ on $(0, 1]$. The function f is continuous on $(0, 1]$ with respect to the Euclidean metrics. We now show that f is not uniformly continuous on $(0, 1]$. For $n \geq 1$, let

$$x_n = \frac{1}{n} \text{ and } y_n = \frac{2}{n}.$$

It is easy to check that the sequence $(d_1(x_n, y_n)) = (|x_n - y_n|)$ converges to 0 while the sequence

$$(d_2(f(x_n), f(y_n))) = (|n - \frac{n}{2}|)$$

is bounded away from 0. Hence, f is not uniformly continuous on $(0, 1]$.

The following important result gives a particular case for which continuity implies uniform continuity.

Theorem 4.2 *Let $f : X \longrightarrow Y$ where (X, d_1) and (Y, d_2) are metric spaces. Let f be continuous on a compact set K in X. Then, f is uniformly continuous on K.*

Proof of Theorem 4.2
We do a proof by contradiction. Assume that f is continuous on the compact K but not uniformly continuous. Then, for some $\epsilon > 0$ there exist two sequences (a_n) and (b_n) in K so that the sequence $(d_1(a_n, b_n))$ converges to 0 and for every $n \geq 1$,

$$d_2(f(a_n), f(b_n)) \geq \epsilon.$$

Since (a_n) is a sequence in a compact set it has a subsequence (a_{n_k}) that converges to some ℓ in K. Observe now that $(d_1(a_{n_k}, b_{n_k}))$ is a subsequence of $(d_1(a_n, b_n))$ which converges to 0. Hence, $(d_1(a_{n_k}, b_{n_k}))$ converges to 0 as well. By the triangle inequality

$$d_1(b_{n_k}, \ell) \leq d_1(b_{n_k}, a_{n_k}) + d_1(a_{n_k}, \ell).$$

Since both terms on the r.h.s. converge to 0 we have that (b_{n_k}) converges to ℓ. By the continuity of f at ℓ, the sequences $(f(a_{n_k}))$ and $(f(b_{n_k}))$ both converge to $f(\ell)$. In particular, $(d_2(f(a_{n_k}), f(b_{n_k})))$ converges to 0. But we also have

$$d_2(f(a_{n_k}), f(b_{n_k})) \geq \epsilon,$$

for every $k \geq 1$. Hence, $(d_2(f(a_{n_k}), f(b_{n_k})))$ converges to 0 and is bounded away from 0! We have a contradiction, f is uniformly continuous on K. This completes the proof of Theorem 4.2.

5 Continuity and Intervals

Unless otherwise specified, continuity in this section refers to continuity with respect to the Euclidean metric on $\mathbb{R}$.

Intervals are the sets of real numbers $(-\infty, +\infty)$, $(-\infty, a)$, (a, b), $(b, +\infty)$ as well as these same sets with or without endpoints. They turn out to be the only sets with the following property.

$(\mathcal{P})$ Let x and y be in S. If $x \leq z \leq y$, then z is in S as well.

Clearly, all intervals have property $(\mathcal{P})$. We now show that the only sets with property $\mathcal{P}$ are intervals.

Proposition 5.1 *Let S be a nonempty subset of $\mathbb{R}$ with property $\mathcal{P}$. Then, S is an interval.*

Proof of Proposition 5.1
Assume that S is bounded above but not bounded below. We will show that S is either $(-\infty, a)$ or $(-\infty, a]$ for some a.

Since S is not empty and is bounded above it has a least upper bound a (why?). In particular, the inclusion $S \subset (-\infty, a]$ holds. We now show the reverse inclusion. Let x be in $(-\infty, a)$. Since a is the least upper bound of S, x is not an upper bound of S. Hence, there exists y in S such that $x < y < a$. Since S is not bounded below there is an $z < x$ that belongs to S. Thus, $z < x < y$ where z and y belong to S. Therefore, by property $\mathcal{P}$, x belongs to S. We have shown the inclusion $(-\infty, a) \subset S$. We already know that $S \subset (-\infty, a]$. So either a belongs to S and $S = (-\infty, a]$ or a does not belong to S and $S = (-\infty, a)$.

All the other cases: S is bounded above and below, S is neither bounded below nor above, S is bounded below but not above, are treated in a similar fashion and are left to the reader.

Theorem 5.1 *Let f be a real valued continuous function on the interval $I \subset \mathbb{R}$. Then, $f(I)$ is an interval.*

Proof of Theorem 5.1
We will show that the set $f(I)$ has property $\mathcal{P}$ and is therefore an interval by Proposition 5.1. Let $x < y$ be in $f(I)$. Thus, there are a and b in I such that $x = f(a)$ and $y = f(b)$. Without loss of generality assume that $a < b$. Let $f(a) < \gamma < f(b)$ and S be the set

$$S = \{x \in [a, b] : f(x) < \gamma\}.$$

Since a belongs to S, S is not empty. Also, S is bounded above by b. Thus, S has a least upper bound c. Using that $a \leq c \leq b$, a and b belong to I and I is an interval we get that c is in I. Therefore, f is continuous at c. Using that c is the least upper bound of S, there exists a sequence (x_n) in S that converges to c. Hence,

$$\lim_n f(x_n) = f(c).$$

The sequence (x_n) is in S so for every $n \geq 1$, $f(x_n) < \gamma$. Taking limits on both sides of the inequality we get

$$f(c) \leq \gamma.$$

Since $\gamma < f(b)$ we see that $c < b$. Thus, $c + 1/n < b$ for n large enough. Since $c + 1/n$ does not belong to S,

$$f(c + \frac{1}{n}) \geq \gamma.$$

Letting n go to infinity we get $f(c) \geq \gamma$. But we already proved that $f(c) \leq \gamma$. Hence, $f(c) = \gamma$. In particular, γ belongs to $f(I)$.

In summary, we have shown that if $f(a) < \gamma < f(b)$, then γ belongs to $f(I)$. Hence, by Proposition 5.1, $f(I)$ is an interval. This completes the proof of Theorem 5.1.

Corollary 5.1 (Intermediate Value Theorem) *Let f be a real valued continuous function on the interval $I \subset \mathbb{R}$. Let a and b in I such that $f(a) < \gamma < f(b)$. Then, there exists a c in I such that $f(c) = \gamma$.*

The proof of Corollary 5.1 is an easy consequence of Theorem 5.1 and is left to the problems.

Theorem 5.2 *Let f be a real valued continuous function on the interval $[a, b] \subset \mathbb{R}$. Then, $f([a, b]) = [\alpha, \beta]$ for some $\alpha < \beta$.*

In words, the continuous image of a closed bounded interval is a closed bounded interval.

Proof of Theorem 5.2
By Theorem 5.1, $f([a, b])$ is an interval. By Theorem 4.1, $f([a, b])$ is a compact set of $\mathbb{R}$. Hence, $f([a, b])$ is a bounded and closed interval (why?). Hence, $f([a, b]) = [\alpha, \beta]$ for some $\alpha < \beta$. This completes the proof of Theorem 5.2.

Corollary 5.2 (Extreme Value Theorem) *Let f be a real valued continuous function on the interval $[a, b] \subset \mathbb{R}$. Then, there exist c and d in $[a, b]$ such that for all x in $[a, b]$ we have*

$$f(c) \leq f(x) \leq f(d).$$

Corollary 5.2 is an immediate consequence of Theorem 5.2. Its proof is left to the problems.

Problems

1. Show that

$$f^{-1}(\bigcap_{i\geq 1} B_i) = \bigcap_{i\geq 1} f^{-1}(B_i).$$

2. Show that

$$\left(f^{-1}(B)\right)^c = f^{-1}(B^c).$$

3. Show that

$$f(A \cup B) = f(A) \cup f(B).$$

4.

(a) Show that

$$f(A \cap B) \subset f(A) \cap f(B).$$

(b) Give an example showing strict inclusion in (a).

5. Let $g : (X, d) \longrightarrow (\mathbb{R}, |.|)$. Assume that g is continuous at a and $g(a) \neq 0$. Let (a_n) be a sequence converging to a. Show that for some N, $g(a_n) \neq 0$ for all $n \geq N$.

6. Give an example of a continuous function f and an open set V such that $f(V)$ is not open.

7. Let f be defined on a metric space (X, d) and have values in $(\mathbb{R}, |.|)$. Assume that f is continuous. Let a be a real number.

(a) Show that the set

$$A = \{x : f(x) > a\}$$

is an open set. Is A a subset of X or $\mathbb{R}$?

(b) Show that the set

$$B = \{x : f(x) \geq a\}$$

is a closed set.

(c) Assume that g is also continuous on X. Let

$$C = \{x : f(x) = g(x)\}.$$

Show that C is closed.

8. Consider the function $p : \mathbb{R}^2 \longrightarrow \mathbb{R}$ defined by $p(x, y) = x$.

(a) Show that p is continuous with respect to the Euclidean metrics.
(b) Describe $p^{-1}(\{0\})$. Is this set closed, open, or neither?

9. Consider the function $g : \mathbb{R}^2 \longrightarrow \mathbb{R}$ defined by

$$g(x, y) = \frac{xy}{x^2 + y^2} \text{ for } (x, y) \neq (0, 0),$$

and $g((0, 0)) = a$ for some real a. Show that the function g is not continuous at $(0, 0)$.

10. Let (X, d_1) and (Y, d_2) be two metric spaces. Let $f : X \longrightarrow Y$ be continuous. Show that for any set A of X we have

$$f\left(\bar{A}\right) \subset \overline{f(A)}.$$

(Recall $\bar{A}$ is the closure of a set A and that an element belongs to $\bar{A}$ if and only if it is the limit of a sequence in A).

11. Assume that the function f is continuous on a compact K and has values in $\mathbb{R}^n$. Show that $f(K)$ is bounded in $\mathbb{R}^n$.

12. Give an example of a continuous function f and a compact set K such that $f^{-1}(K)$ is not compact.

13. Give an example showing that the inverse image of an interval by a continuous function need not be an interval.

14. Prove Corollary 5.1.

15. Prove Corollary 5.2.

16. Give an example of a continuous function f and an open interval I such that $f(I)$ is not open.

17. Let $P : \mathbb{R} \longrightarrow \mathbb{R}$ be a nonconstant polynomial. In this problem we show that $|P|$ attains its minimum.

(a) Show that the range of $|P|$ has a greatest lower bound m.
(b) Show that there exists a sequence (x_n) such that $(|P(x_n)|)$ converges to m.
(c) Show that there exists $a > 0$ such that if $|x| > a$, then $|P(x)| > m + 1$.
(d) Show that there exists N such that if $n \geq N$, then $|x_n| \leq a$.
(e) Show that there exists an ℓ in $[-a, a]$ such that $|P(\ell)| = m$.

18. Let f be a continuous function on the interval I with integer values. Show that f is a constant function.

Chapter 10
Measurable Sets and Measurable Functions

The Riemann integral has multiple issues. In particular, too many functions are not Riemann integrable, interchange of limit and integral requires stringent hypotheses, and the Riemann integral does not provide a complete metric for continuous functions. This is why we need a more general integration theory. This theory was discovered by Henri Lebesgue in the early 1900s. This chapter and the next do the ground work for the Lebesgue integral. The integral will be introduced in Chapter 12.

The first step (done in the first section) is to decide which sets of the real line can be measured, they will be called *measurable* sets. It turns out that the set of real numbers has subsets that are too pathological to be measured, this is why the notion of measurable sets is critical to the theory. The second step (done in the second section) is to define *measurable functions*, i.e., the functions on which we can apply our new integral.

The main motivation for measure theory is the Lebesgue measure on the real line. However, the theory is no more difficult (and may be easier to understand) if we do it on a general space with a general measure. This is the approach we take.

1 Measurable Sets

Definition 1.1 *Let X be a nonempty set. A* ***σ-algebra*** *$\mathcal{A}$ on X is a nonempty collection of sets with the following properties*

(i) *If $E_1, E_2, \ldots$ belong to $\mathcal{A}$ so does the countable union $\bigcup_{i=1}^{\infty} E_i$.*
(ii) *If E belongs to $\mathcal{A}$ so does its complement E^c.*

In words, a σ-algebra on X is a nonempty collection of subsets which is closed under taking complements and countable unions.

R. B. Schinazi, *From Classical to Modern Analysis*,
https://doi.org/10.1007/978-3-319-94583-5_10

The symbol σ is used to emphasize that we may take unions over a finite or an *infinite* (but countable) collection of sets.

- If E belongs to a σ-algebra $\mathcal{A}$, then E is said to be **measurable** with respect to $\mathcal{A}$.

  We now turn to important consequences of this definition. Let $\mathcal{A}$ be a σ-algebra on some set X.
- If $E_1, E_2, \ldots$ belong to $\mathcal{A}$, so does the intersection $\bigcap_{i=1}^{\infty} E_i$. That is, a σ-algebra is closed under countable intersections.

  We now prove this property. By (ii), E_i^c belongs to $\mathcal{A}$ for every $i \geq 1$. By (i), we have that $\cup_{i=1}^{\infty} E_i^c$ belongs to $\mathcal{A}$. By (ii) the complement

$$\left(\bigcup_{i=1}^{\infty} E_i^c\right)^c = \bigcap_{i=1}^{\infty} E_i$$

  belongs to $\mathcal{A}$. This proves the property.
- The empty set $\emptyset$ belongs to $\mathcal{A}$.

  Since $\mathcal{A}$ is not empty there is some E in $\mathcal{A}$. Hence, E^c belongs to $\mathcal{A}$ as well and so does $E \cap E^c = \emptyset$.
- The whole set X belongs to $\mathcal{A}$.

  Take E in $\mathcal{A}$. Then E^c belongs to $\mathcal{A}$ as well and so does $E \cup E^c = X$.

  We now turn to examples of σ-algebras.

Example 1.1 $\mathcal{P}(X)$ (the collection of all subsets of X) is a σ-algebra on X.

Example 1.2 Let $\mathcal{A} = \{\emptyset, X\}$. $\mathcal{A}$ is a σ-algebra on X.

The claims in Examples 1.1 and 1.2 are easy to prove and are left as exercises. These examples give trivial σ-algebras. To get the nontrivial ones the following property will be important.

Proposition 1.1 *The intersection over any nonempty collection of σ-algebras on X is a σ-algebra.*

Note that the collection in Proposition 1.1 need not be countable.

Proof of Proposition 1.1
Let $(\mathcal{A}_i)_{i \in I}$ be a collection of σ-algebras on X. Let

$$\mathcal{A} = \bigcap_{i \in I} \mathcal{A}_i.$$

As noted above $\emptyset$ and X belong to every σ-algebra on X. Hence, $\mathcal{A}$ is a nonempty collection of subsets of X.

Let $E_1, E_2, \ldots$ be in $\mathcal{A}$. Then for every $i \in I$ we have $\bigcup_{j \geq 1} E_j \in \mathcal{A}_i$ (why?) and therefore $\bigcup_{j \geq 1} E_j \in \mathcal{A}$.

Let E be in $\mathcal{A}$. Then for every $i \in I$ we have $E^c \in \mathcal{A}_i$ (why?) and therefore $E^c \in \mathcal{A}$.

This completes the proof of Proposition 1.1.

Proposition 1.2 *Let $\mathcal{F}$ be a collection of sets in X. There exists a unique smallest σ-algebra $m(\mathcal{F})$ that contains $\mathcal{F}$. That is, if $\mathcal{F} \subset \mathcal{A}$ where $\mathcal{A}$ is a σ-algebra, then $m(\mathcal{F}) \subset \mathcal{A}$.*

The σ-algebra $m(\mathcal{F})$ is said to be the σ-algebra **generated** by $\mathcal{F}$.

Proof of Proposition 1.2
Consider the collection of all the σ-algebras containing $\mathcal{F}$. Note that $\mathcal{P}(X)$ is such a σ-algebra since it contains every subset of X. Thus, this collection of σ-algebras is not empty. Let $m(\mathcal{F})$ be the intersection over all the σ-algebras containing $\mathcal{F}$. By Proposition 1.1, $m(\mathcal{F})$ is a σ-algebra.

We now show that $m(\mathcal{F})$ is the smallest σ-algebra containing $\mathcal{F}$. Let $\mathcal{A}$ be a σ-algebra containing $\mathcal{F}$. Hence, $\mathcal{A}$ is part of the collection of σ-algebras containing $\mathcal{F}$. The intersection over this collection of σ-algebras is precisely $m(\mathcal{F})$. Therefore, $m(\mathcal{F}) \subset \mathcal{A}$. This completes the proof of Proposition 1.2.

We are now ready to apply the preceding propositions to find a nontrivial σ-algebra.

Consider the collection of all open sets (with respect to the Euclidean metric) on the real line. The σ-algebra generated by this collection of sets is called the **Borel** σ-algebra and will be denoted by $\mathcal{B}$.

- Open sets on the real line are Borel measurable. A closed set is the complement of an open set. Thus, closed sets are also Borel measurable.

Proposition 1.3 *The Borel σ-algebra $\mathcal{B}$ is generated by the collection of all open bounded intervals*

$$\mathcal{F} = \{(a, b) : a < b\}.$$

That is, $m(\mathcal{F}) = \mathcal{B}$.

- There is nothing special about the collection of intervals $\{(a, b) : a < b\}$. The Borel σ-algebra is also generated by $\{[a, b) : a < b\}$, $\{(-\infty, b) : b \in \mathbb{R}\}$ and so on.

Proposition 1.3 shows that a σ-algebra that contains all open bounded intervals must necessarily contain all open sets of $\mathbb{R}$. Before proving Proposition 1.3 we show that $m(\mathcal{F})$ contains all the open intervals, including the unbounded ones.

Example 1.3 Show that all open intervals $(-\infty, a)$ belong to $m(\mathcal{F})$.

It is enough to show that

$$(-\infty, a) = \bigcup_{n \geq 1} (-n, a) \tag{1.1}$$

As always we prove the equality of two sets by showing a double inclusion. For every $n \geq 1$ we have $(-n, a) \subset (-\infty, a)$. Hence,

$$\bigcup_{n \geq 1} (-n, a) \subset (-\infty, a).$$

For the reverse inclusion, assume $a > 0$. Take x in $(-\infty, a)$. If $x \geq 0$, then x is in $(-1, a)$. If $x < 0$ there exists a natural n_0 such that $n_0 > -x$ (by the Archimedean Property). Therefore, x belongs to $(-n_0, a)$. In either case there is an $n \geq 1$ such that x belongs to $(-n, a)$. Therefore,

$$(-\infty, a) \subset \bigcup_{n \geq 1} (-n, a).$$

This proves (1.1) for $a > 0$. For $a \leq 0$ the proof is even simpler and is left as an exercise.

Example 1.4 Show that all open intervals $(a, +\infty)$ belong to $m(\mathcal{F})$.

Observe that the complement of $(a, +\infty)$ is $(-\infty, a]$ and that

$$(-\infty, a] = (-\infty, a) \cup \{a\}.$$

By Example 1.3 we know that $(-\infty, a)$ belongs to $m(\mathcal{F})$. Hence, if we can show that $\{a\}$ belongs to $m(\mathcal{F})$, then $(-\infty, a]$ is in $m(\mathcal{F})$ and so is $(a, +\infty)$. It is enough to show that

$$\{a\} = \bigcap_{n \geq 1} (a - \frac{1}{n}, a + \frac{1}{n}). \tag{1.2}$$

Since $\{a\}$ is included in $(a - \frac{1}{n}, a + \frac{1}{n})$ for every n, it is included in the intersection over all these intervals. This proves one inclusion. For the reverse inclusion, let x be in $\bigcap_{n \geq 1} (a - \frac{1}{n}, a + \frac{1}{n})$. For every $n \geq 1$, we have

$$a - \frac{1}{n} < x < a + \frac{1}{n}.$$

We now let n go to infinity and we get, $a \leq x \leq a$. That is, $x = a$. This completes the proof of (1.2).

Proof of Proposition 1.3
Since open intervals are open sets we have that $\mathcal{F} \subset \mathcal{B}$ and hence by Proposition 1.2 that $m(\mathcal{F}) \subset \mathcal{B}$.

To show the reverse inclusion we need the following about open sets in $\mathbb{R}$.

Lemma 1.1 *Every open set in $\mathbb{R}$ equipped with the Euclidean metric is a countable union of open intervals.*

Outline of the proof of Lemma 1.1
Let U be an open set in $\mathbb{R}$. Let x be in U. Since U is open we can find an open ball $I_x \subset U$. In the Euclidean topology an open ball is an open interval. Hence, we have

$$U \subset \bigcup_{x \in U} I_x.$$

Since every I_x is a subset of U we also have the reverse inclusion. This shows that U is a union of open intervals. Note that this union is over all x in U and U need not be countable. The difficulty in this proof is to show that the union can be taken to be countable. For a proof of this fact see for instance Krantz (1991).

We are now ready to show that $\mathcal{B} \subset m(\mathcal{F})$. Let $\mathcal{U}$ be the collection of all open sets in $\mathbb{R}$.

By Lemma 1.1 every open set of $\mathbb{R}$ is in $m(\mathcal{F})$ (why?). That is, $\mathcal{U} \subset m(\mathcal{F})$ Hence, by Proposition 1.2, we have $m(\mathcal{U}) \subset m(\mathcal{F})$. Since $m(\mathcal{U}) = \mathcal{B}$, the proof of Proposition 1.3 is complete.

As we will see in the next example and in the problems the Borel σ-algebra $\mathcal{B}$ can be generated in a number of ways.

Example 1.5 Let

$$\mathcal{F}_1 = \{(-\infty, a) : a \in \mathbb{R}\}.$$

Show that $m(\mathcal{F}_1) = \mathcal{B}$.

Let $\mathcal{F} = \{(a, b) : a < b\}$. By Example 1.3, all intervals $(-\infty, a)$ are in $m(\mathcal{F})$. Thus, $\mathcal{F}_1 \subset m(\mathcal{F})$ and by Proposition 1.2, $m(\mathcal{F}_1) \subset m(\mathcal{F})$.

We now prove the reverse inclusion. For $a < b$ we have

$$(a, b) = (a, +\infty) \cap (-\infty, b).$$

If we show that $(a, +\infty)$ belongs to $m(\mathcal{F}_1)$, then we will have that (a, b) belongs to $m(\mathcal{F}_1)$ as well. We have

$$(-\infty, a] = \bigcap_{n \geq 1} (-\infty, a + \frac{1}{n}) \tag{1.3}$$

We now prove (1.3). Clearly, for every $n \geq 1$ we have $(-\infty, a] \subset (-\infty, a + \frac{1}{n})$. Hence,

$$(-\infty, a] \subset \bigcap_{n \geq 1} (-\infty, a + \frac{1}{n}).$$

For the reverse inclusion take x in $\bigcap_{n \geq 1} (-\infty, a + \frac{1}{n})$. For every $n \geq 1$ we have

$$x < a + \frac{1}{n}.$$

Letting n go to infinity yields $x \leq a$. Thus, x is in $(-\infty, a]$. This completes the proof of (1.3).

By (1.3), $(-\infty, a]$ is in $m(\mathcal{F}_1)$. Hence, its complement $(a, +\infty)$ is also in that σ-algebra. This implies that $(a, b) = (a, +\infty) \cap (-\infty, b)$ is in $m(\mathcal{F}_1)$. Therefore, $\mathcal{F} \subset m(\mathcal{F}_1)$. By Proposition 1.2 we have $m(\mathcal{F}) \subset m(\mathcal{F}_1)$. This completes the proof that $m(\mathcal{F}_1) = m(\mathcal{F}) = \mathcal{B}$.

- Arguments such as the one above show that many subsets of the reals are Borel sets. Intersections of open and closed sets are Borel sets. Intersections and unions of the resulting sets are also Borel and so on. However, not every subset of the reals is Borel! For a discussion of these issues see Folland (1999).

Problems

1. Show that $\mathcal{P}(X)$ is a σ-algebra on X.

2. Let $\mathcal{A} = \{\emptyset, X\}$.

(a) Show that $\mathcal{A}$ is a σ-algebra on X.
(b) Let $\mathcal{B}$ be a σ-algebra on X. Show that $\mathcal{A} \subset \mathcal{B}$.

3. Consider the Borel σ-algebra $\mathcal{B}$.

(a) Show that a countable intersection of open sets is in $\mathcal{B}$.
(b) Show that a countable union of closed sets is in $\mathcal{B}$.

4. Recall that the Cantor set $\mathcal{C}$ is a subset of the reals

$$\mathcal{C} = \bigcap_{n\geq} S_n$$

where each S_n is a union of closed intervals. Show that $\mathcal{C}$ is in the Borel σ-algebra.

5. Let $\mathcal{F}_2 = \{[a, b] : a < b\}$.

(a) Show that every $[a, b]$ belongs to $m(\mathcal{F})$ where $\mathcal{F} = \{(a, b) : a < b\}$.
(b) Use (a) to show that $m(\mathcal{F}_2) \subset m(\mathcal{F})$.
(c) Show that for $a < b$ we have

$$(a, b) = \bigcup_{n\geq 1} [a + \frac{1}{n}, b - \frac{1}{n}].$$

(d) Use (c) to show that every (a, b) belongs to $m(\mathcal{F}_2)$.
(e) Conclude that $\mathcal{F}_2$ generates the Borel σ-algebra.

6. Let $\mathcal{F}_3 = \{(a, b] : a < b\}$. Show that $\mathcal{F}_3$ generates the Borel σ-algebra. (Follow the pattern of Problem 6..)

7. Let $\mathcal{F}_4 = \{[a, +\infty) : a \in \mathbb{R}\}$. Show that $\mathcal{F}_4$ generates the Borel σ-algebra.

8. Consider the Borel σ-algebra $\mathcal{B}$.

(a) Let x be a real number. Show that $\{x\}$ is in $\mathcal{B}$.
(b) Show that the set of rational numbers is in $\mathcal{B}$.
(c) Show that the set of irrational numbers is in $\mathcal{B}$.

2 Measurable Functions

2.1 Definition and General Properties

Recall that if $f : X \longrightarrow Y$ and if $E \subset Y$, then $f^{-1}(E)$ is defined as

$$f^{-1}(E) = \{x \in X : f(x) \in E\}.$$

Recall also that

$$f^{-1}(E^c) = \left(f^{-1}(E)\right)^c,$$

and that for any sequence of subsets $E_1, E_2, \ldots$ of Y we have

$$f^{-1}\left(\bigcup_{n\geq 1} E_n\right) = \bigcup_{n\geq 1} f^{-1}(E_n),$$

$$f^{-1}\left(\bigcap_{n\geq 1} E_n\right) = \bigcap_{n\geq 1} f^{-1}(E_n).$$

An application of these properties gives the following useful result.

Proposition 2.1 *Let $f : X \longrightarrow Y$ and let $\mathcal{M}$ be a σ-algebra on X. Define*

$$\mathcal{A} = \{E \subset Y : f^{-1}(E) \in \mathcal{M}\}.$$

Then, $\mathcal{A}$ is a σ-algebra on Y.

Proof of Proposition 2.1
Note that $\mathcal{A}$ is not empty since $f^{-1}(Y) = X$ and $X \in \mathcal{M}$. Hence, $Y \in \mathcal{A}$.

Let E be in $\mathcal{A}$. Then, $f^{-1}(E)$ is in $\mathcal{M}$ and therefore $\left(f^{-1}(E)\right)^c$ is also in $\mathcal{M}$ (why?). Since

$$f^{-1}(E^c) = \left(f^{-1}(E)\right)^c,$$

$f^{-1}(E^c)$ is in $\mathcal{M}$ as well. That is, if E belongs to $\mathcal{A}$ so does E^c.

Consider $E_1, E_2, \ldots,$ in $\mathcal{A}$. Then, $f^{-1}(E_1), f^{-1}(E_2), \ldots$ are all in $\mathcal{M}$ and so is $\bigcup_{n\geq 1} f^{-1}(E_n)$ (why?). Since

$$f^{-1}\left(\bigcup_{n\geq 1} E_n\right) = \bigcup_{n\geq 1} f^{-1}(E_n),$$

$f^{-1}\left(\bigcup_{n\geq 1} E_n\right)$ is in $\mathcal{M}$ as well. Therefore, $\bigcup_{n\geq 1} E_n$ is in $\mathcal{A}$. This completes the proof of Proposition 2.1.

Definition 2.1 *Let $f : X \longrightarrow Y$, and let $\mathcal{M}$ and $\mathcal{N}$ be σ-algebras on X and Y, respectively. The function f is said to be* ***measurable*** *with respect to $\mathcal{M}$ and $\mathcal{N}$ if for every E in $\mathcal{N}$ we have that $f^{-1}(E)$ is in $\mathcal{M}$.*

We will just say that f is measurable when there is no ambiguity on what the σ-algebras are. For many application $Y = \mathbb{R}$ and $\mathcal{N} = \mathcal{B}$, the Borel σ-algebra.

The following criterion for measurability will be quite useful.

Proposition 2.2 *Let $f : X \longrightarrow Y$, and let $\mathcal{M}$ and $\mathcal{N}$ be σ-algebras on X and Y, respectively. Assume that $\mathcal{F}$ is a collection of subsets of Y and that it generates $\mathcal{N}$, i.e., $m(\mathcal{F}) = \mathcal{N}$. Then, f is measurable if and only if $f^{-1}(E)$ is in $\mathcal{M}$ for every E in $\mathcal{F}$.*

Proof of Proposition 2.2

Assuming that f is measurable we have that $f^{-1}(E)$ is in $\mathcal{M}$ for every E in $\mathcal{F}$ since $\mathcal{F} \subset \mathcal{N}$. This proves the direct implication. For the converse consider

$$\mathcal{A} = \{E \subset Y : f^{-1}(E) \in \mathcal{M}\}.$$

By Proposition 2.1 $\mathcal{A}$ is a σ-algebra on Y. We also have that $\mathcal{F} \subset \mathcal{A}$. Therefore, by Proposition 1.2, $m(\mathcal{F}) \subset \mathcal{A}$. That is, for every E in $m(\mathcal{F}) = \mathcal{N}$ we have $f^{-1}(E) \in \mathcal{M}$. This proves that f is measurable with respect to $\mathcal{M}$ and $\mathcal{N}$. This completes the proof of Proposition 2.2.

2.2 Real Valued Functions

Unless otherwise specified the real line will always be assumed to be equipped with the Borel σ-algebra $\mathcal{B}$ (i.e., the σ-algebra generated by the open sets of $\mathbb{R}$). Assume that $\mathcal{M}$ is a σ-algebra on X. Then a function $f : X \longrightarrow \mathbb{R}$ will be said to be $\mathcal{M}$ measurable if it is measurable with respect to $\mathcal{M}$ and $\mathcal{B}$. We will say that a function is Borel measurable in the particular case where $X = \mathbb{R}$ and $\mathcal{M} = \mathcal{B}$. When there is no ambiguity about which σ-algebras are used we will just say that f is measurable.

Proposition 2.3 *Let $f : \mathbb{R} \longrightarrow \mathbb{R}$ be a continuous function, then f is Borel measurable.*

Proof of Proposition 2.3
Let U be an open set of $\mathbb{R}$. Since f is continuous $f^{-1}(U)$ is open and hence belongs to $\mathcal{B}$. Since the open sets generate $\mathcal{B}$, by Proposition 2.2 f is Borel measurable. This completes the proof of Proposition 2.3.

Next we give a convenient way to check that a real valued function is measurable. We will use the notation

$$\{f < a\} \text{ for } \{x \in X : f(x) < a\}.$$

Proposition 2.4 *Let $\mathcal{M}$ be a σ-algebra on X and consider the function $f : X \longrightarrow \mathbb{R}$. Then f is $\mathcal{M}$ measurable if and only for every real a the set $\{f < a\}$ belongs to $\mathcal{M}$.*

- There is nothing special about the sets $\{f < a\}$. Proposition 2.4 holds as well for sets $\{f \leq a\}$, $\{f \geq a\}$ or $\{f > a\}$.

Proof of Proposition 2.4
Assume that f is measurable. Since open sets belong to the Borel σ-algebra, $f^{-1}((-\infty, a)) = \{f < a\}$ belongs to $\mathcal{M}$ for every real a. This proves the direct implication.

For the converse, note that by Example 1.3, $\mathcal{F}_1 = \{(-\infty, a) : a \in \mathbb{R}\}$ generates the Borel σ-algebra. By hypothesis, $f^{-1}((-\infty, a))$ is in $\mathcal{M}$ for every a. Hence, by Proposition 2.2 f is $\mathcal{M}$ measurable. The proof of Proposition 2.4 is complete.

As the next example shows a measurable function need not be continuous. First a notation, for a set A the function $\mathbf{1}_A$ is called the **indicator function** of A and is defined as

$$\mathbf{1}_A(x) = 1 \text{ if } x \in A$$

$$\mathbf{1}_A(x) = 0 \text{ if } x \notin A$$

Example 2.1 Consider the Dirichlet function $f = \mathbf{1}_{\mathbb{Q}} : \mathbb{R} \longrightarrow \mathbb{R}$. It is not difficult to check that this function is nowhere continuous. Show that f is Borel measurable.

Here, $f = X \longrightarrow Y$ where $X = Y = \mathbb{R}$ and we are taking $\mathcal{B}$ as the sigma-algebra for both X and Y.

Let $a \leq 0$. Then $\{f < a\} = \emptyset$ which is measurable with respect to any σ-algebra.

Let $a > 1$. Then $\{f < a\} = \mathbb{R}$ which is in $\mathcal{B}$.

Let $0 < a \leq 1$. Then $\{f < a\} = \mathbb{Q}^c$ the set of irrational numbers (why?). We will show that $\mathbb{Q}$ is in $\mathcal{B}$ and this will imply that $\mathbb{Q}^c$ is also in $\mathcal{B}$. Recall that the set of rational numbers is countable. That is, we can write $\mathbb{Q} = \{r_1, r_2, \ldots\}$, where (r_n) is a sequence. Hence, $\mathbb{Q}$ is the following countable union

$$\mathbb{Q} = \bigcup_{n \geq 1} \{r_n\}$$

and each singleton $\{r_n\}$ is in $\mathcal{B}$ (why?). Hence, $\mathbb{Q}$ is in $\mathcal{B}$. This completes the proof that f is Borel measurable.

2.3 Operations on Measurable Functions

Proposition 2.5 *Let $\mathcal{M}$ be a σ-algebra on X. Consider the functions $f : X \longrightarrow \mathbb{R}$ and $g : \mathbb{R} \longrightarrow \mathbb{R}$. Assume that f is $\mathcal{M}$ measurable and g is continuous, then $g \circ f$ is $\mathcal{M}$ measurable.*

Proof of Proposition 2.5
Let U be an open set in $\mathbb{R}$. Then $(g \circ f)^{-1}(U) = f^{-1}\left(g^{-1}(U)\right)$ (see the problems). By the continuity of g, $g^{-1}(U)$ is open and is therefore a Borel set. Since f is $\mathcal{M}$ measurable, then $f^{-1}\left(g^{-1}(U)\right)$ is in $\mathcal{M}$. Note now that the collection of open sets generates the Borel σ-algebra. Hence, by Proposition 2.2, $g \circ f$ is $\mathcal{M}$ measurable. This completes the proof of Proposition 2.5.

Proposition 2.6 *Let $\mathcal{M}$ be a σ-algebra on X. Consider the functions $f : X \longrightarrow \mathbb{R}$ and $g : X \longrightarrow \mathbb{R}$. Assume that f and g are $\mathcal{M}$ measurable. Then,*

(i) *$f + g$ is $\mathcal{M}$ measurable.*
(ii) *fg is $\mathcal{M}$ measurable.*
(iii) *If $g \neq 0$ on X, then $\frac{f}{g}$ is $\mathcal{M}$ measurable.*

Proof of Proposition 2.6
We first prove (i). Let $r_1, r_2, \ldots$ be an enumeration of the rational numbers. Then,

$$\{f + g < a\} = \{f < a - g\} = \bigcup_{n \geq 1} \{f < r_n < a - g\},$$

where the last equality comes from the density of the rationals in the reals (i.e., if $x < y$, there exists a rational r_n such that $x < r_n < y$). Note now that for every $n \geq 1$

$$\{f < r_n < a - g\} = \{f < r_n\} \cap \{a - g > r_n\}.$$

Both sets on the r.h.s. are in $\mathcal{M}$ (why?) and therefore so is the set $\{f < r_n < a - g\}$. A countable union of measurable sets is measurable. Therefore, for every a we have that $\{f + g < a\}$ is measurable. This proves that $f + g$ is $\mathcal{M}$ measurable and completes the proof of (i).

We now turn to (ii). First note that if g is measurable so is cg for any constant c (why?). Hence, $(-g)$ is measurable and by (i) so is $f - g$. By Proposition 2.5, $(f + g)^2$ and $(f - g)^2$ are both measurable. Hence, so is

$$(f + g)^2 - (f - g)^2 = 4fg,$$

and therefore fg. This completes the proof of (ii).

(iii) is an easy consequence of Proposition 2.5 and (ii) and is left as an exercise. This completes the proof of Proposition 2.6.

2.4 Sequences of Measurable Functions

Let (a_n) be a sequence of real numbers. Recall that $\sup_{n\geq 1} a_n$ is the least upper bound of the set $\{a_n : n \in \mathbb{N}\}$, provided the sequence is bounded above. If the sequence is not bounded above we let $\sup_{n\geq 1} a_n = +\infty$.

Let (f_n) be a sequence of real valued functions defined on X. The function $\sup_n f_n$ is defined by $\sup_n f_n(x)$ for every x in X. Because of our convention on unbounded real sequences the function $\sup_n f_n$ may take infinite values. The function $\inf_n f_n$ is defined in analogous way and is also allowed to take infinite values as well. We define $\limsup_n f_n$ and $\liminf f_n$ by

$$\limsup_n f_n(x) = \inf_{n\geq 1} g_n(x), \quad \liminf_n f_n(x) = \sup_{n\geq 1} h_n(x)$$

where $g_n(x) = \sup_{k\geq n} f_k(x)$ and $h_n(x) = \inf_{k\geq n} f_k(x)$.

Proposition 2.7 *Let $\mathcal{M}$ be a σ-algebra on X. Consider the sequence (f_n) of functions $f_n : X \longrightarrow \mathbb{R}$. Assume that for all $n \geq 1$, f_n is $\mathcal{M}$ measurable. Then,*

(i) $\sup_{n\geq 1} f_n$ *is $\mathcal{M}$ measurable.*
(ii) $\inf_{n\geq 1} f_n$ *is $\mathcal{M}$ measurable.*
(iii) $\limsup_n f_n$ *is $\mathcal{M}$ measurable.*
(iv) $\liminf_n f_n$ *is $\mathcal{M}$ measurable.*

Proof of Proposition 2.7
Let a be a real number. For any $x \in X$, $\sup_{n\geq 1} f_n(x) \leq a$ if and only if $f_n(x) \leq a$ for every $n \geq 1$. Hence,

$$\{\sup_{n\geq 1} f_n \leq a\} = \bigcap_{n\geq 1} \{f_n \leq a\}.$$

Since $\{f_n \leq a\}$ is in $\mathcal{M}$ so is $\{\sup_{n\geq 1} f_n \leq a\}$. It is not difficult to see that the collection of intervals

$$\{(-\infty, a] : a \in \mathbb{R}\}$$

generates the Borel σ-algebra, see Section 1. Hence, by Proposition 2.2, $\sup_{n\geq 1} f_n$ is $\mathcal{M}$ measurable. This completes the proof of (i).

The proof of (ii) is quite similar. Observe that

$$\{\inf_{n\geq 1} f_n \geq a\} = \bigcap_{n\geq 1}\{f_n \geq a\}.$$

The collection of intervals

$$\{([a, +\infty) : a \in \mathbb{R}\}$$

generates the Borel σ-algebra and we apply again Proposition 2.2. The details are left to the reader.

We now turn to (iii). Recall that

$$\limsup_n f_n(x) = \inf_{n\geq 1} g_n(x),$$

where $g_n(x) = \sup_{k\geq n} f_k(x)$. By (ii) every g_n is measurable. Therefore, by (i) $\limsup_n f_n$ is measurable. This proves (iii).

The proof of (iv) is very similar to the proof of (iii) and is left to the reader. The proof of Proposition 2.7 is complete.

Corollary 2.1 *Let $\mathcal{M}$ be a σ-algebra on X. Consider a sequence (f_n) of real valued measurable functions that converges pointwise to some f. Then, f is measurable.*

Proof of Corollary 2.1
Let x in X. Since the sequence of real numbers $(f_n(x))$ converges to $f(x)$ we have that

$$f(x) = \limsup_n f_n(x).$$

Hence, $f = \limsup_n f_n$ which is measurable by Proposition 2.7. This completes the proof of Corollary 2.1.

2.5 Simple Functions

Consider a set X equipped with a σ-algebra $\mathcal{M}$. Let n be a natural number, $a_1, a_2, \ldots, a_n$ be distinct positive real numbers, and $A_1, \ldots, A_n$ be $\mathcal{M}$ measurable sets. We define the function $s : X \longrightarrow [0, +\infty)$ by

$$s(x) = a_i \text{ if and only if } x \in A_i.$$

It is convenient to write s using indicator functions,

$$s = \sum_{i=1}^{n} a_i \mathbf{1}_{A_i}.$$

Such a function is called a **simple function**. It is not difficult to see that s is $\mathcal{M}$ measurable (see the Problems).

The following result will be critical in the construction of the Lebesgue integral.

Theorem 2.1 *Let $f : X \longrightarrow [0, +\infty]$ be $\mathcal{M}$ measurable. There exists a sequence (s_n) of simple functions such that for every x the sequence $(s_n(x))$ is an increasing sequence (i.e., $s_1(x) \leq s_2(x) \leq \dots$) that converges to $f(x)$.*

In words, f is the pointwise limit of an increasing sequence of simple functions. Note that we allow f to be infinite.

Proof of Theorem 2.1
Let $n \geq 1$ be a natural number. Let $y \geq 0$ be a real number. Any real number is between two consecutive integers. Hence, there exists a positive integer $k_n(y)$ such that

$$k_n(y) \leq 2^n y < k_n(y) + 1 \tag{2.1}$$

Therefore,

$$2^{-n} k_n(y) \leq y < 2^{-n} k_n(y) + 2^{-n} \tag{2.2}$$

We define the function $g_n : [0, +\infty] \longrightarrow [0, +\infty)$ by

$$g_n(y) = 2^{-n} k_n(y) \text{ if } 0 \leq y < n$$

$$g_n(y) = n \text{ if } n \leq y \leq +\infty$$

Note that g_n is defined on the positive reals and at $+\infty$. For $k = 0, 1, \dots, n2^n - 1$ let

$$I(k, n) = [2^{-n} k, 2^{-n} k + 2^{-n}).$$

Observe that these intervals are disjoint and that their union is $[0, n)$. Hence, g_n can be written as

$$g_n = \sum_{k=0}^{n2^n - 1} 2^{-n} k_n(y) \mathbf{1}_{I(k,n)} + n \mathbf{1}_{[n,+\infty]}.$$

All intervals are Borel measurable, therefore g_n is Borel measurable. Note also that g_n is a simple function.

We now show that for every y in $[0, +\infty]$ the sequence $(g_n(y))$ is increasing and converges to y. There are two cases to consider.

- If $y < +\infty$, there exists a natural $N > y$ (why?). Using (2.2) we have for $n \geq N$

$$y - 2^{-n} < g_n(y) \leq y.$$

  Hence, $(g_n(y))$ converges to y.

  By (2.1) we have $2k_n(y) \leq 2^{n+1}y$. Note that $k_{n+1}(y)$ is the largest positive integer which is smaller than $2^{n+1}y$. Hence, $k_{n+1}(y) \geq 2k_n(y)$ and $g_{n+1}(y) \geq g_n(y)$. That is, the sequence $(g_n(y))$ is increasing.
- If $y = +\infty$, then $g_n(y) = n$ for every n. Hence, $(g_n(y))$ increases to y as n goes to infinity.

Let $s_n = g_n \circ f$. For every $n \geq 1$, s_n is a simple function. This is due to the fact that $\mathbf{1}_A \circ f = \mathbf{1}_B$ where $B = f^{-1}(A)$ (why?). Since g_n is Borel measurable s_n is $\mathcal{M}$ measurable (why?).

Using the fact that for every y, $(g_n(y))$ is increasing and converges to y we get that for every x in X, $(s_n(x))$ is increasing and converges to $f(x)$. This completes the proof of Theorem 2.1.

Problems

1. Assume that X is equipped with the σ-algebra $\mathcal{P}(X)$ and Y with a σ-algebra $\mathcal{A}$. Show that every function $f : X \longrightarrow Y$ is measurable.

2. Assume that X is equipped with the σ-algebra $\mathcal{A} = \{\emptyset, X\}$ and $\mathbb{R}$ with the Borel σ-algebra. Show that a function $f : X \longrightarrow \mathbb{R}$ is measurable if and only if it is constant.

3. Show that the Dirichlet function $\mathbf{1}_{\mathbb{Q}}$ is nowhere continuous.

4. Consider the indicator function $\mathbf{1}_A : X \longrightarrow \mathbb{R}$ where A is a subset of X. Let $\mathcal{M}$ be a σ-algebra on X. Show that $\mathbf{1}_A$ is measurable if and only if A belongs to $\mathcal{M}$.

5. Consider the functions $f : X \longrightarrow Y$ and $g : Y \longrightarrow Z$. Let $U \subset Z$. Show that $(g \circ f)^{-1}(U) = f^{-1}\left(g^{-1}(U)\right)$.

6. Consider the functions $f : X \longrightarrow \mathbb{R}$ and $g : X \longrightarrow \mathbb{R}$. Assume that f and g are measurable with respect to $\mathcal{M}$. Show that the set $\{f < g\}$ is measurable.

7. Show that a constant function $f : X \longrightarrow \mathbb{R}$ is measurable.

8. Show that if $g : X \longrightarrow \mathbb{R}$ is measurable so is cg for any constant c.

9. Prove Proposition 2.6 (iii).

10. Prove Proposition 2.7 (iv).

11. Let (f_n) be a sequence of real valued measurable functions. Consider the set

$$A = \{x \in X : (f_n(x)) \text{ converges.}\}.$$

Show that A is a measurable set.

12. Let f be a real valued measurable function. Let $f^+ = \max(f, 0)$. Show that f^+ is measurable.

13. Consider a set X equipped with a σ-algebra $\mathcal{M}$. Assume that $A_1, \ldots, A_n$ are subsets of X such that $A_i \cap A_j = \emptyset$ for $i \neq j$. We define the function $s : X \longrightarrow [0, +\infty)$ by

$$s = \sum_{i=1}^{n} a_i \mathbf{1}_{A_i}.$$

Show that s is measurable if and only if $A_1, \ldots, A_n$ are measurable.

14. Consider a set X equipped with a σ-algebra $\mathcal{M}$. Let $f : X \longrightarrow [0, +\infty]$ be $\mathcal{M}$ measurable.

(a) Let A be a Borel set. Show that $\mathbf{1}_A \circ f = \mathbf{1}_B$ where $B = f^{-1}(A)$.
(b) Let $s : [0, +\infty)] \longrightarrow [0, +\infty)$ be a simple Borel function. Show that $s \circ f$ is an $\mathcal{M}$ measurable simple function.

Chapter 11
Measures

1 Definition and Examples

In Chapter 10 we defined the notion of measurable sets. We now turn to the task of actually measuring these sets. The main example of a measure on the real line is the so-called Lebesgue measure for which the measure of an interval is its length. There are, however, many other ways to measure a set. In this chapter we define and study measures.

Definition 1.1 *Let $\mathcal{M}$ be a σ-algebra on a set X. A measure μ on $\mathcal{M}$ is a function*

$$\mu : \mathcal{M} \longrightarrow [0, +\infty]$$

with the following properties

(i) $\mu(\emptyset) = 0$.
(ii) *μ is countably additive. That is, if $E_1, E_2, \ldots$ is a sequence of disjoint sets in $\mathcal{M}$, then*

$$\mu\left(\bigcup_{n=1}^{\infty} E_n\right) = \sum_{n=1}^{\infty} \mu(E_n).$$

In words, a measure is a function that assigns a positive real value (possibly $+\infty$) to a measurable set.

The triplet $(X, \mathcal{M}, \mu)$ will be called a **measure space**.

Observe that (ii) implies finite additivity (why?). That is, if $E_1, E_2, \ldots E_n$ is a finite collection of disjoint sets in $\mathcal{M}$, then

$$\mu\left(\bigcup_{k=1}^{n} E_k\right) = \sum_{k=1}^{n} \mu(E_k).$$

R. B. Schinazi, *From Classical to Modern Analysis*,
https://doi.org/10.1007/978-3-319-94583-5_11

Example 1.1 Consider the set $\mathbb{N}$ of natural numbers equipped with the σ-algebra $\mathcal{P}(\mathbb{N})$. Let $f : \mathbb{N} \longrightarrow [0, +\infty)$. Define the function μ by

$$\mu(A) = \sum_{x \in A} f(x),$$

where $A \subset \mathbb{N}$. We show now that μ is a measure on $\mathcal{P}(\mathbb{N})$.

By convention a sum over an empty set of indices is 0. Hence, (i) in Definition 1.1 holds.

Recall that positive terms can be rearranged in any order in a series without changing the sum, see Theorem 5.2 in Chapter 4. Hence, we do not need to be more precise as how the sum is computed in $\mu(A)$. We now prove countable additivity. Let $E_1, E_2, \ldots$ be a sequence of disjoint sets in $\mathcal{P}(\mathbb{N})$. Let

$$E = \bigcup_{n=1}^{\infty} E_n.$$

Since $E_1, E_2, \ldots$ are disjoint and partition E, one of the ways to sum the series over E is to sum over each E_n for $n \geq 1$. Hence,

$$\mu(E) = \sum_{x \in E} f(x) = \sum_{n \geq 1} \sum_{x \in E_n} f(x).$$

Therefore,

$$\mu\left(\bigcup_{n=1}^{\infty} E_n\right) = \sum_{n=1}^{\infty} \mu(E_n).$$

This proves countable additivity for μ. Thus, μ is a measure on $\mathcal{P}(\mathbb{N})$.

In the particular case where f is identically 1, μ is called the **counting measure**.

We now turn to several important properties of measures.

- We will use the notation $B \backslash A$ for $B \cap A^c$.

Proposition 1.1 *Let $(X, \mathcal{M}, \mu)$ be a measure space.*

(a) *Let A and B be in $\mathcal{M}$. If $A \subset B$, then $\mu(A) \leq \mu(B)$.*
(b) *Let A and B be in $\mathcal{M}$. Assume that $A \subset B$ and $\mu(B) < +\infty$. Then,*

$$\mu(B \backslash A) = \mu(A) - \mu(B).$$

(c) *Let $E_1, E_2, \ldots$ be any sequence in $\mathcal{M}$, then*

$$\mu\left(\bigcup_{n=1}^{\infty} E_n\right) \leq \sum_{n=1}^{\infty} \mu(E_n).$$

Since μ can take infinite values, one or both sides of the inequalities in (a) and (c) may be infinite.

Proof of Proposition 1.1
We start with (a). Note that $B \cap A$ and $B \cap A^c$ are disjoint and that their union is B. Hence,

$$\mu(B) = \mu(B \cap A) + \mu(B \cap A^c).$$

Since $A \subset B$, we have $B \cap A = A$. Thus,

$$\mu(B) = \mu(A) + \mu(B \cap A^c) \tag{1.1}$$

In particular, $\mu(A) \leq \mu(B)$. This proves (a).

We now turn to (b). It is an easy consequence of (1.1). Since $\mu(B)$ is finite so is $\mu(A)$ (why?). Hence, by (1.1)

$$\mu(B) - \mu(A) = \mu(B \cap A^c) = \mu(B \backslash A).$$

This proves (b). Note that the assumption $\mu(B) < \infty$ is necessary. Otherwise, we could end up with $\mu(B) - \mu(A) = +\infty - \infty$ which is undefined.

We now prove (c). Let $F_1 = E_1$, $F_2 = E_1 \backslash E_2$, $F_3 = E_3 \backslash (E_1 \cup E_2)$ and so on. More precisely, for $n \geq 2$ we set

$$F_n = E_n \backslash \left(\bigcup_{i=1}^{n-1} E_i\right).$$

Note that for every $n \geq 1$, $F_n \subset E_n$, the sets $F_1, F_2, \ldots$ are disjoint and that

$$\bigcup_{n=1}^{\infty} E_n = \bigcup_{n=1}^{\infty} F_n.$$

Hence,

$$\mu\left(\bigcup_{n=1}^{\infty} E_n\right) = \mu\left(\bigcup_{n=1}^{\infty} F_n\right) = \sum_{n=1}^{\infty} \mu(F_n).$$

By part (a) of Proposition 1.1, $\mu(F_n) \leq \mu(E_n)$ for every $n \geq 1$. Therefore,

$$\mu\left(\bigcup_{n=1}^{\infty} E_n\right) \leq \sum_{n=1}^{\infty} \mu(E_n).$$

This completes the proof of (c) and of Proposition 1.1.

Proposition 1.2 *Let $(X, \mathcal{M}, \mu)$ be a measure space.*

(a) *Let $E_1, E_2, \ldots$ be an increasing sequence of sets in $\mathcal{M}$. That is, for every $n \geq 1$ we have $E_n \subset E_{n+1}$. Then,*

$$\lim_{n\to\infty} \mu(E_n) = \mu\left(\bigcup_{n=1}^{\infty} E_n\right).$$

(b) *Let $E_1, E_2, \ldots$ be a decreasing sequence of sets in $\mathcal{M}$. That is, for every $n \geq 1$ we have $E_{n+1} \subset E_n$. Assuming also that $\mu(E_1) < \infty$, then*

$$\lim_{n\to\infty} \mu(E_n) = \mu\left(\bigcap_{n=1}^{\infty} E_n\right).$$

Before giving the proof of Proposition 1.2 we give an example showing that (b) need not be true without the assumption $\mu(E_1) < \infty$.

Example 1.2 Consider the measure space $(\mathbb{N}, \mathcal{P}(\mathbb{N}), \nu)$ where ν is the counting measure, see Example 1.1. For $n \geq 1$, let E_n be all the natural numbers larger than or equal to n. Then, $\mu(E_n) = +\infty$ for every $n \geq 1$. Hence,

$$\lim_{n\to\infty} \mu(E_n) = +\infty.$$

However, $\bigcap_{n=1}^{\infty} E_n = \emptyset$ (why?). Hence,

$$\mu(\bigcap_{n=1}^{\infty} E_n) = 0.$$

Therefore, Proposition 1.2 (b) does not hold in this case.

Proof of Proposition 1.2
We first prove (a). There are two cases to consider.

- If there exists $N \geq 1$ such that $\mu(E_N) = +\infty$, then $\mu(E_k) = +\infty$ for every $k \geq N$ (why?). Then,

$$\lim_{n\to\infty} \mu(E_n) = +\infty.$$

Since

$$\mu\left(\bigcup_{n=1}^{\infty} E_n\right) \geq \mu(E_N) = +\infty,$$

we have

$$\lim_{n\to\infty} \mu(E_n) = \mu\left(\bigcup_{n=1}^{\infty} E_n\right).$$

This proves (a) in this case.

- We now assume that $\mu(E_n) < +\infty$ for every $n \geq 1$. Let $F_1 = E_1$ and $F_n = E_n \backslash E_{n-1}$ for $n \geq 2$.

Using that (E_n) is an increasing sequence of sets we have that the sets F_n are disjoint and

$$\bigcup_{n=1}^{\infty} E_n = \bigcup_{n=1}^{\infty} F_n.$$

Hence,

$$\mu\left(\bigcup_{n=1}^{\infty} E_n\right) = \mu\left(\bigcup_{n=1}^{\infty} F_n\right) = \sum_{n=1}^{\infty} \mu(F_n).$$

Note that $\mu(F_1) = \mu(E_1)$ and by Proposition 1.1 (b), $\mu(F_n) = \mu(E_n) - \mu(E_{n-1})$ for every $n \geq 2$. Hence,

$$\sum_{k=1}^{n} \mu(F_k) = \mu(E_1) + \sum_{k=2}^{n} (\mu(E_k) - \mu(E_{k-1})) = \mu(E_n).$$

Observe now that the l.h.s. has a (possibly infinite) limit since all the terms in the sum are positive. We let n go to infinity to get

$$\sum_{n=1}^{\infty} \mu(F_n) = \lim_{n\to\infty} \mu(E_n),$$

where both sides may be infinite. This completes the proof of (a).

We now turn to (b). Let $G_i = E_1 \backslash E_i$, for $i = 1, 2, \ldots$. Since the sequence (E_i) is decreasing, the sequence (G_i) is increasing (why?). We apply Proposition 1.2 (a) to get

$$\lim_{n\to\infty} \mu(G_n) = \mu\left(\bigcup_{n=1}^{\infty} G_n\right) \tag{1.2}$$

Observe now that,

$$\bigcup_{n=1}^{\infty} G_n = \bigcup_{n=1}^{\infty} (E_1 \backslash E_n) = \bigcup_{n=1}^{\infty} \left(E_1 \cap E_n^c\right).$$

Using the distributivity of union over intersection we get

$$\bigcup_{n=1}^{\infty} G_n = E_1 \bigcap \left(\bigcup_{n=1}^{\infty} E_n^c\right) = E_1 \bigcap \left(\bigcap_{n=1}^{\infty} E_n\right)^c = E_1 \backslash \left(\bigcap_{n=1}^{\infty} E_n\right).$$

Given that $\mu(E_1) < \infty$ and $E_n \subset E_1$, Proposition 1.1 (b) applies, hence

$$\mu(G_n) = \mu(E_1) - \mu(E_n). \tag{1.3}$$

By the same argument,

$$\mu\left(\bigcup_{n=1}^{\infty} G_n\right) = \mu\left(E_1 \backslash \left(\bigcap_{n=1}^{\infty} E_n\right)\right) = \mu(E_1) - \mu\left(\bigcap_{n=1}^{\infty} E_n\right) \tag{1.4}$$

We use (1.3) and (1.4) in (1.2) to get

$$\lim_{n\to\infty} (\mu(E_1) - \mu(E_n)) = \mu(E_1) - \mu\left(\bigcap_{n=1}^{\infty} E_n\right).$$

Since $\mu(E_1) < \infty$ we can cancel it on both sides of the last equality to get Proposition 1.2 (b). This completes the proof of the proposition.

Next we give an application of Proposition 3.2 which is quite useful in probability theory. First, some notation. Let (A_n) be a sequence of sets in a measure space $(X, \mathcal{M}, \mu)$. Let

$$\limsup A_n = \bigcap_{n\geq 1} \bigcup_{k\geq n} A_k.$$

The notation is analogous to the one used for real number sequences but note that $\limsup A_n$ is a set. Moreover, $\limsup A_n$ is in $\mathcal{M}$ (why?). This set has an interesting interpretation. It is easy to see that x belongs to $\limsup A_n$ if and only if x belongs to A_n for infinitely many $n \geq 1$ (see the Problems).

Theorem 1.1 (Borel-Cantelli Lemma) *Let $(X, \mathcal{M}, \mu)$ be a measure space and (A_n) be a sequence of measurable sets. Assume that*

$$\sum_{n\geq 1} \mu(A_n) < +\infty.$$

Then, $\mu(\limsup A_n) = 0$.

Proof of Theorem 1.1
For $n \geq 1$, let

$$E_n = \bigcup_{k \geq n} A_k.$$

Note that (E_n) is a decreasing sequence in $\mathcal{M}$ (why?). Observe also that

$$\mu(E_1) \leq \sum_{n \geq 1} \mu(A_n) < +\infty.$$

Hence, by Proposition 1.2 we have

$$\lim_{n \to \infty} \mu(E_n) = \mu\left(\bigcap_{n \geq 1} E_n\right).$$

Note that the r.h.s. is precisely $\mu(\limsup A_n)$. To finish the proof we only need to show that $\lim_{n \to \infty} \mu(E_n) = 0$. By Proposition 1.1,

$$\mu(E_n) = \mu\left(\bigcup_{k \geq n} A_k\right) \leq \sum_{k \geq n} \mu(A_k).$$

Since the series $\sum_{n \geq 1} \mu(A_n)$ converges, we have

$$\lim_{n \to +\infty} \sum_{k \geq n} \mu(A_k) = 0.$$

Hence, the sequence $(\mu(E_n))$ also converges to 0. The proof of Theorem 1.1 is complete.

2 The Lebesgue Measure on the Real Line

The **Lebesgue measure** m is a measure on the real line with the following property. For any interval (a, b),

$$m\left((a, b)\right) = b - a.$$

Is there such a measure? It turns out that there is such a measure. However, the proof of the existence of the Lebesgue measure is rather technical and we omit it. In this section, we will state several results without proving them, see Folland (1999) for proofs.

The Lebesgue measure cannot be defined for all the subsets of $\mathbb{R}$. Some sets in $\mathbb{R}$ are just too pathological to be measured and need to be excluded. This is why we need the whole measurability theory of Chapter 10.

Theorem 2.1 *Let $F : \mathbb{R} \longrightarrow \mathbb{R}$ be an increasing, right continuous function. There exists a unique measure μ_F defined on the Borel σ-algebra $\mathcal{B}$ such that*

$$\mu_F((a, b]) = F(b) - F(a).$$

Theorem 2.1 provides the existence of the Lebesgue measure m. We simply take $F(x) = x$ to get the existence of a measure m with the following property. For any $a < b$

$$m((a, b]) = F(b) - F(a) = b - a.$$

It is convenient to define m on a larger σ-algebra that includes the Borel σ-algebra $\mathcal{B}$. Let $\mathcal{L}$ be the **Lebesgue σ-algebra**. It includes $\mathcal{B}$ and has the following additional property. If A is in $\mathcal{L}$ and $m(A) = 0$, then any subset of A is also in $\mathcal{L}$. Because of this property the Lebesgue measure space is said to be **complete**.

Moreover, the Lebesgue measure m has the following important properties.

- The Lebesgue measure is translation invariant. That is, if A is in $\mathcal{L}$ and x is a real, then

$$m(A + x) = m(A),$$

  where $A + x = \{a + x; a \in A\}$.
- If A is in $\mathcal{L}$ and x is a real, then

$$m(xA) = |x|m(A),$$

  where $xA = \{xa; a \in A\}$.

2.1 Null Sets

A **null set** N in a measure space $(X, \mathcal{M}, \mu)$ is such that N belongs to $\mathcal{M}$ and

$$\mu(N) = 0.$$

The empty set is always a null set. In general, there are many other null sets. Next, we give some examples on the Lebesgue measure space.

- A singleton is a null set.

  To prove the claim let a be a real number and let $A_n = (a - \frac{1}{n}, a + \frac{1}{n}]$ for every $n \geq 1$. Since a belongs to A_n for every $n \geq 1$ we have

$$m(\{a\}) \leq m(A_n).$$

  We have for every $n \geq 1$,

$$m(A_n) = a + \frac{1}{n} - (a + \frac{1}{n}) = \frac{2}{n},$$

  and therefore

$$m(\{a\}) \leq \frac{2}{n}.$$

  We let n go to infinity to get $m(\{a\}) = 0$.
- Since the Lebesgue measure of a singleton is 0 we have for every $a < b$ that

$$m\left([a, b]\right) = m\left((a, b]\right) = m\left([a, b)\right) = m\left((a, b)\right) = b - a.$$

- Any countable set in $\mathbb{R}$ is a null set.

  Let $S = \{a_1, a_2, \dots\}$ be a countable subset of real numbers (such as the set of rationals). Since

$$S = \bigcup_{i=1}^{\infty} \{a_i\}$$

  and each singleton $\{a_i\}$ is a Borel set (why?) so is S. Hence, S is in $\mathcal{L}$. Moreover,

$$m(S) \leq \sum_{i=1}^{\infty} m(\{a_i\}) = 0.$$

  Hence, $m(S) = 0$.

  Are there uncountable sets whose Lebesgue measure is 0? The answer is Yes.
- The Cantor set is an uncountable null set.

We first recall the definition of the Cantor set. Let $S_0 = [0, 1]$. We remove the middle third of S_0 to get $S_1 = [0, \frac{1}{3}] \cup [\frac{2}{3}, 1]$. Then, we remove the middle third of both intervals in S_1 to get S_2 and so on. For $n \geq 0$, S_n is the union of 2^n disjoint intervals of length (i.e., Lebesgue measure) 3^{-n}. Hence, for $n \geq 0$

$$m(S_n) = \frac{2^n}{3^n}.$$

The Cantor set is defined as

$$\mathcal{C} = \bigcap_{n \geq 1} S_n.$$

Hence, for every $n \geq 1$, $\mathcal{C} \subset S_n$. Thus,

$$m(\mathcal{C}) \leq m(S_n) = \frac{2^n}{3^n}.$$

We let n go to infinity and we get $m(\mathcal{C}) = 0$. The Cantor set is a null set.

2.2 Approximation of Lebesgue Measurable Sets and Functions

While Lebesgue measurable sets and functions can be quite intricate, we will see next that they can be approximated by much simpler objects.

The next property is the so-called **First Littlewood's Principle**: a Lebesgue measurable set is almost a union of open intervals.

To precisely state the principle we need the following notation. The symmetric difference $A \Delta B$ of two sets A and B is defined by

$$A \Delta B = (A \cap B^c) \cup (A^c \cap B).$$

In words, $A \Delta B$ is the set of elements that are in A but not in B or in B but not in A.

Theorem 2.2 *Let E be Lebesgue measurable. Assume that $m(E) < +\infty$. For any $\epsilon > 0$ there exists a set U which is a countable union of open intervals such that*

$$m(E \Delta U) < \epsilon.$$

We now turn to measurable functions.

Theorem 2.3 (Lusin's Theorem) Let $f : [a, b] \longrightarrow \mathbb{R}$ be a Lebesgue measurable function. For any $\epsilon > 0$ there is a compact subset K of $[a, b]$ on which f is continuous and such that $m(K^c) < \epsilon$.

In words, a Lebesgue measurable function is almost continuous! This is the **Second Littlewood's Principle**.

Problems

In the problems below we will assume that $(X, \mathcal{M}, \mu)$ is a measure space.

1. Show that (ii) in Definition 1.1 implies finite additivity. That is, if $E_1, E_2, \dots E_n$ are disjoint sets in $\mathcal{M}$, then

$$\mu\left(\bigcup_{k=1}^{n} E_k\right) = \sum_{k=1}^{n} \mu(E_k).$$

2. Consider an increasing sequence (E_n) in $\mathcal{M}$.

(a) Show that if $\mu(E_n) = +\infty$ for some n, then $\mu(E_k) = +\infty$ for every $k \geq n$.
(b) Show that if $\mu(E_n) = +\infty$ for some $n \geq 1$, then $\mu\left(\bigcup_{n=1}^{\infty} E_n\right) = +\infty$.

3. Consider a decreasing sequence (E_n) in $\mathcal{M}$. Let $G_n = E_1 \backslash E_n$, for $i = 1, 2, \dots$. Show that the sequence (G_n) is increasing.

4. Consider a sequence (A_n) in $\mathcal{M}$. Let

$$\limsup A_n = \bigcap_{n\geq 1} \bigcup_{k\geq n} A_k.$$

(a) Show that $\limsup A_n$ belongs to $\mathcal{M}$.
(b) Show that x belongs to $\limsup A_n$ if and only if x belongs to A_n for infinitely many $n \geq 1$.

5. Consider a sequence (A_n) in $\mathcal{M}$. Let

$$\liminf A_n = \bigcup_{n\geq 1} \bigcap_{k\geq n} A_k.$$

(a) Show that $\liminf A_n$ belongs to $\mathcal{M}$.
(b) Give a characterization of $\liminf A_n$ analogous to the one in 4 (b).

6. Show that for any A and B in $\mathcal{M}$ we have

$$\mu(A \cup B) + \mu(A \cap B) = \mu(A) + \mu(B).$$

(Write $A \cup B$ as the union of two disjoint sets.)

7. Consider a decreasing sequence (A_n) in $\mathcal{M}$. Assume that $\bigcap_{n\geq 1} A_n = \emptyset$ and that $\mu(A_1) < \infty$. Show that

$$\lim_{n\to\infty} \mu(A_n) = 0.$$

8. Consider the Lebesgue measure space. Assume that N is in $\mathcal{L}$ and that N is a null set. Show that a subset of N is also a null set.

9. Show that $m(\mathbb{R}) = +\infty$, where m is the Lebesgue measure.

10. Assume that μ is a measure on the Borel σ-algebra $\mathcal{B}$. Assume also that $\mu(\mathbb{R}) < +\infty$. We define the function $F : \mathbb{R} \longrightarrow \mathbb{R}$ by

$$F(x) = \mu\left((-\infty, x]\right).$$

(a) Show that F is increasing.
(b) Show that F is right continuous.

11. Let (E_n) be a sequence of measurable sets in $(X, \mathcal{M}, \mu)$.

(a) Show that for every $n \geq 1$

$$\mu\left(\bigcap_{k \geq n} E_k\right) \leq \mu(E_n).$$

(b) Use (a) to show that

$$\mu\left(\liminf E_n\right) \leq \liminf_n \mu(E_n).$$

12. Let (E_n) be a sequence of measurable sets in $(X, \mathcal{M}, \mu)$. Assume that $\mu\left(\bigcup_{k\geq 1} E_k\right)$ is finite.

(a) Show that

$$\mu\left(\limsup E_n\right) \geq \limsup_n \mu(E_n).$$

(b) Assume that there is $a > 0$ such that $\mu(E_n) > a$ for infinitely many $n \geq 1$. Show that $\mu\left(\limsup E_n\right) > 0$.
(c) Show that the inequality in (a) need not hold if $\mu\left(\bigcup_{k\geq 1} E_k\right)$ is infinite.

13. Let $\psi : \mathcal{M} \longrightarrow [0, +\infty)$ be a **finite values** function with the following two properties.

(i) The function ψ is finitely additive (i.e., if $A_1, \ldots, A_n$ are disjoint, then $\psi\left(\bigcup_{k=1}^n A_k\right) = \sum_{k=1}^n \psi(A_k)$).
(ii) If (A_n) is a decreasing sequence in $\mathcal{M}$ such that $\bigcap_{n\geq 1} A_n = \emptyset$, then $\lim_{n\to\infty} \psi(A_n) = 0$. We will show that ψ is a measure on $\mathcal{M}$.

  (a) Show that $\psi(\emptyset) = 0$. (Observe that $A = A \cup \emptyset$).
  (b) Assume that $A \subset B$ are in $\mathcal{M}$. Show that $\psi(B\backslash A) = \psi(B) - \psi(A)$.
  The next step is to show that ψ is countably additive. Let (E_n) be an infinite sequence of disjoint sets in $\mathcal{M}$. Let E be the union of these sets. For every $n \geq 1$ let

$$A_n = E \setminus \left(\bigcup_{k=1}^{n} E_k \right).$$

(c) Show that (A_n) is a decreasing sequence of sets.
(d) Show that $\bigcap_{k=1}^{\infty} A_k = \emptyset$.
(e) Show that for every $n \geq 1$

$$\psi(E) = \psi \left(\bigcup_{k=1}^{n} E_k \right) + \psi(A_n).$$

(f) Use (e) to show that

$$\psi(E) = \sum_{k=1}^{\infty} \psi(E_k).$$

Chapter 12
The Lebesgue Integral

We construct the Lebesgue integral in three steps. First for simple functions, then for positive measurable functions, and finally for all real valued measurable functions. The more technical work is done for simple functions and is then used for the more general sets of functions.

1 The Integral of a Positive Function

In this section we fix a measure space $(X, \mathcal{M}, \mu)$. Measurability refers to this space unless otherwise specified.

1.1 The Integral of a Simple Function

Recall that a simple measurable function s can be represented as

$$s = \sum_{i=1}^{n} a_i \mathbf{1}_{A_i},$$

where n is a natural number, $a_1, a_2, \ldots, a_n$ are real numbers and $A_1, A_2, \ldots, A_n$ belong to $\mathcal{M}$. In this section we will consider only positive simple functions. Hence, $a_i \geq 0$ for $i = 1, 2, \ldots, n$. Moreover, we assume that $A_1, A_2, \ldots, A_n$ form a **partition** of X. That is, $A_1, A_2, \ldots, A_n$ are disjoint and

$$\bigcup_{i=1}^{n} A_i = X.$$

R. B. Schinazi, *From Classical to Modern Analysis*,
https://doi.org/10.1007/978-3-319-94583-5_12

We define the **Lebesgue integral** of s with respect to μ as

$$\int s d\mu = \sum_{i=1}^{n} a_i \mu(A_i).$$

In the preceding sum we may have $a_i = 0$ and $\mu(A_i) = +\infty$ for some i. In this case we set $a_i \mu(A_i) = 0$. In particular if s is identically 0, then we have

$$\int 0 d\mu = 0\mu(X) = 0,$$

whether $\mu(X)$ is finite or infinite.

Note that the representation of s is not unique (why not?). Therefore we need to check that our definition of the integral is the same for any representation of s. To do so assume that

$$s = \sum_{i=1}^{n} a_i \mathbf{1}_{A_i} = \sum_{j=1}^{p} b_j \mathbf{1}_{B_j},$$

where $A_1, A_2, \ldots, A_n$ and $B_1, B_2, \ldots, B_p$ are both partitions of X. We need to make sure that

$$\sum_{i=1}^{n} a_i \mu(A_i) = \sum_{j=1}^{p} b_j \mu(B_j).$$

Since $B_1, B_2, \ldots, B_p$ is a partition of X, we have for every i

$$A_i = \bigcup_{j=1}^{p} \left(A_i \cap B_j\right),$$

where the union is over disjoint sets (why?). Using that μ is additive,

$$\mu(A_i) = \sum_{j=1}^{p} \mu(A_i \cap B_j).$$

Hence,

$$\sum_{i=1}^{n} a_i \mu(A_i) = \sum_{i=1}^{n} \sum_{j=1}^{p} a_i \mu(A_i \cap B_j).$$

Similarly, we have

$$\sum_{j=1}^{p} b_j \mu(B_j) = \sum_{j=1}^{p} \sum_{i=1}^{n} b_j \mu(A_i \cap B_j).$$

If $A_i \cap B_j \neq \emptyset$ for some i and j, then there is at least one x in $A_i \cap B_j$. We have $s(x) = a_i$ and $s(x) = b_j$. Hence, if $A_i \cap B_j \neq \emptyset$, then $a_i = b_j$. On the other hand, if $A_i \cap B_j = \emptyset$, then $\mu(A_i \cap B_j) = 0$. Therefore, for every i and j we have

$$a_i \mu(A_i \cap B_j) = b_j \mu(A_i \cap B_j).$$

Since the double sums above are finite we are allowed to invert the order of summation. Thus,

$$\sum_{i=1}^{n} a_i \mu(A_i) = \sum_{j=1}^{p} b_j \mu(B_j).$$

This proves that our definition of $\int s d\mu$ does not depend on the representation of s.

Example 1.1 Consider the indicator function $\mathbf{1}_{\mathbb{Q}}$ of the rational numbers. Recall that this function is not Riemann integrable. We now show that it has a Lebesgue integral.

We know that the set of rationals $\mathbb{Q}$ is Lebesgue measurable and is a null set. Let m be the Lebesgue measure on $\mathbb{R}$. We have

$$\mathbf{1}_{\mathbb{Q}} = 1 \times \mathbf{1}_{\mathbb{Q}} + 0 \times \mathbf{1}_{\mathbb{Q}^c}.$$

By definition of the integral of a simple function we have

$$\int \mathbf{1}_{\mathbb{Q}} dm = m(\mathbb{Q}).$$

Since $m(\mathbb{Q}) = 0$ we have that

$$\int \mathbf{1}_{\mathbb{Q}} dm = 0.$$

We now turn to the first properties of the Lebesgue integral.

Proposition 1.1 *Let s and t be positive measurable simple functions.*

(a) $\int cs d\mu = c \int s d\mu$ *for any real* $c \geq 0$.
(b) $\int (s + t) d\mu = \int s d\mu + \int t d\mu$.
(c) If $s \leq t$, *then* $\int s d\mu \leq \int t d\mu$.

Proof of Proposition 1.1
The proof of (a) is easy and is left to the reader. We turn to (b).

Let $s = \sum_{i=1}^{n} a_i \mathbf{1}_{A_i}$ and $t = \sum_{j=1}^{p} b_j \mathbf{1}_{B_j}$, where $A_1, A_2, \ldots, A_n$ and $B_1, B_2, \ldots, B_p$ are both partitions of X. We have by definition of the integral that

$$\int s d\mu + \int t d\mu = \sum_{i=1}^{n} a_i \mu(A_i) + \sum_{j=1}^{p} b_j \mu(B_j).$$

Using that $A_1, A_2, \ldots, A_n$ and $B_1, B_2, \ldots, B_p$ are partitions of X we have

$$\sum_{i=1}^{n} a_i \mu(A_i) = \sum_{i=1}^{n} a_i \sum_{j=1}^{p} \mu(A_i \cap B_j),$$

and

$$\sum_{j=1}^{p} b_j \mu(B_j) = \sum_{j=1}^{p} b_j \sum_{i=1}^{n} \mu(A_i \cap B_j).$$

Hence,

$$\begin{aligned}\int s d\mu + \int t d\mu &= \sum_{i=1}^{n} a_i \sum_{j=1}^{p} \mu(A_i \cap B_j) + \sum_{j=1}^{p} b_j \sum_{i=1}^{n} \mu(A_i \cap B_j)\\ &= \sum_{i=1}^{n} \sum_{j=1}^{p} (a_i + b_j) \mu(A_i \cap B_j).\end{aligned}$$

We now compute $\int (s+t) d\mu$. Observe that for every i and j we have

$$\mathbf{1}_{A_i} = \sum_{j=1}^{p} \mathbf{1}_{A_i \cap B_j} \text{ and } \mathbf{1}_{B_j} = \sum_{i=1}^{n} \mathbf{1}_{A_i \cap B_j}.$$

Hence,

$$s + t = \sum_{i=1}^{n} a_i \mathbf{1}_{A_i} + \sum_{j=1}^{p} b_j \mathbf{1}_{B_j} = \sum_{i=1}^{n} \sum_{j=1}^{p} (a_i + b_j) \mathbf{1}_{A_i \cap B_j}.$$

Moreover, the collection of sets $(A_i \cap B_j)$ for $i = 1, \ldots, n$ and $j = 1, \ldots, p$ is a partition of X (why?). Hence, by the definition of the integral we have

$$\int (s+t) d\mu = \sum_{i=1}^{n} \sum_{j=1}^{p} (a_i + b_j) \mu(A_i \cap B_j).$$

This completes the proof of (b).

We now prove (c). Note that if $A_i \cap B_j \neq \emptyset$, then $s(x) = a_i$ and $t(x) = b_j$ for x in $A_i \cap B_j$. Since $s(x) \leq t(x)$ we have $a_i \leq b_j$. On the other hand, if $A_i \cap B_j = \emptyset$, then $\mu(A_i \cap B_j) = 0$. Hence, for all i and j we have

$$a_i \mu(A_i \cap B_j) \leq b_j \mu(A_i \cap B_j).$$

Therefore,

$$\sum_{i=1}^{n} \sum_{j=1}^{p} a_i \mu(A_i \cap B_j) \leq \sum_{i=1}^{n} \sum_{j=1}^{p} b_j \mu(A_i \cap B_j). \tag{1.1}$$

Using that

$$\sum_{j=1}^{p} \mu(A_i \cap B_j) = \mu(A_i) \text{ and } \sum_{i=1}^{n} \mu(A_i \cap B_j) = \mu(B_j),$$

in (1.1) we get

$$\sum_{i=1}^{n} a_i \mu(A_i) \leq \sum_{j=1}^{p} b_j \mu(B_j).$$

That is,

$$\int s d\mu \leq \int t d\mu.$$

This completes the proof of Proposition 1.1.

The next proposition gives a way to construct new measures.

Proposition 1.2 *Let s be a positive measurable simple function. For A in $\mathcal{M}$ let*

$$\nu(A) = \int s \mathbf{1}_A d\mu.$$

Then, ν is a measure on $\mathcal{M}$.

- We will also use the notation $\int_A s d\mu$ for $\int s \mathbf{1}_A d\mu$.

Proof of Proposition 1.2
Note that $\mathbf{1}_\emptyset$ is identically 0. Hence, $s\mathbf{1}_\emptyset$ is also identically 0 and

$$\nu(\emptyset) = \int 0 d\mu = 0.$$

Next we show that ν is countably additive. Let (E_n) be a sequence of disjoint measurable sets and let $E = \bigcup_{n\geq 1} E_n$.

Let $s = \sum_{i=1}^{p} a_i \mathbf{1}_{A_i}$ where p is a natural number and $A_1, A_2, \ldots, A_p$ is a partition of X. Observe that

$$s\mathbf{1}_E = \sum_{i=1}^{p} a_i \mathbf{1}_{E\cap A_i}.$$

Hence, $s\mathbf{1}_E$ is a positive simple function and by the definition of the integral we have

$$\nu(E) = \int s\mathbf{1}_E d\mu = \sum_{i=1}^{p} a_i \mu(E \cap A_i).$$

For every fixed $i = 1, \ldots, p$, $(E_n \cap A_i)$ is a sequence of disjoint measurable sets. Since μ is a measure

$$\mu(E \cap A_i) = \mu\left(\bigcup_{n\geq 1} E_n \cap A_i\right) = \sum_{n\geq 1} \mu(E_n \cap A_i).$$

Thus,

$$\int s\mathbf{1}_E d\mu = \sum_{i=1}^{p} a_i \sum_{n\geq 1} \mu(E_n \cap A_i).$$

Therefore,

$$\int s\mathbf{1}_E d\mu = \sum_{n\geq 1} \sum_{i=1}^{p} a_i \mu(E_n \cap A_i),$$

where we interchanged the sums. The interchange is allowed since a finite sum of limits is the limit of the sum. See the problems. Observe now that for every $n \geq 1$ we have

$$\sum_{i=1}^{p} a_i \mu(E_n \cap A_i) = \int s\mathbf{1}_{E_n} d\mu = \nu(E_n).$$

Hence,

$$\nu(E) = \int s\mathbf{1}_E d\mu = \sum_{n\geq 1} \nu(E_n).$$

That is, ν is a measure on $\mathcal{M}$. The proof of Proposition 1.2 is complete.

Problems

1. Assume that $B_1, B_2, \ldots, B_p$ is a measurable partition of X. Let A be measurable.

(a) Show that

$$\mu(A) = \sum_{i=1}^{p} \mu(A \cap B_i).$$

(b) Show that

$$\mathbf{1}_A = \sum_{j=1}^{p} \mathbf{1}_{A \cap B_j}.$$

2. Let $A_1, A_2, \ldots, A_n$ and $B_1, B_2, \ldots, B_p$ be two measurable partitions of X. Let $a_1, \ldots, a_n$ and $b_1, \ldots, b_p$ such that

$$\sum_{i=1}^{n} a_i \mathbf{1}_{A_i} = \sum_{j=1}^{p} b_j \mathbf{1}_{B_j}.$$

Show that for all i and j we have

$$a_i \mu(A_i \cap B_j) = b_j \mu(A_i \cap B_j).$$

3. Let $A_1, A_2, \ldots, A_n$ and $B_1, B_2, \ldots, B_p$ be two measurable partitions of X. Show that the collection of sets $(A_i \cap B_j)$ for $i = 1, \ldots, n$ and $j = 1, \ldots, p$ is a partition of X.

4. Prove Proposition 1.1 (a).

5.

(a) For every fixed $i = 1, \ldots, p$ assume that $(b_k(i))$ is a positive increasing sequence. That is, $b_k(i) \leq b_{k+1}(i)$ for every i and k. Show that

$$\lim_{k \to \infty} \sum_{i=1}^{p} b_k(i) = \sum_{i=1}^{p} \lim_{k \to \infty} b_k(i).$$

(b) Use (a) to show that

$$\sum_{n\geq 1}\sum_{i=1}^{p} a_i\mu(E_n\cap A_i)=\sum_{i=1}^{p} a_i\sum_{n\geq 1}\mu(E_n\cap A_i),$$

where the $a_i \geq 0$ are real numbers, (A_i) and (E_n) are measurable sets. (Let $b_k(i)=a_i\sum_{n=1}^{k}\mu(E_n\cap A_i)$.)

1.2 The Integral of a Positive Real Function

Let f be a positive measurable function which is allowed to take infinite values. Consider the set

$$A(f)=\{s \text{ simple } : 0\leq s\leq f\}.$$

In other words, $A(f)$ is the collection of positive simple functions that are smaller than f. We define the Lebesgue integral of f as

$$\int f d\mu = \sup\left\{\int s d\mu : s\in A(f)\right\}.$$

As always, if the set is not bounded above we take the sup to be $+\infty$. In this case, we set $\int f d\mu = +\infty$.

If f is a positive simple function, we have two definitions for the integral of f! One definition is the definition of the integral of a simple function and the other is the sup definition. We need to check that both definitions yield the same result. Let a be the integral of f using the simple function definition. Let b be the integral of f using the sup definition. Note that if $f\geq 0$ is a simple function, then f belongs to $A(f)$. Moreover, if s belongs to $A(f)$, then

$$0\leq s\leq f.$$

By Proposition 1.1 (c) we have for any s in $A(f)$

$$\int s d\mu \leq \int f d\mu = a.$$

Hence, a is an upper bound of $\left\{\int s d\mu : s\in A(f)\right\}$. Moreover, a belongs to this set (why?). Hence,

$$\sup\left\{\int s d\mu : s\in A(f)\right\}=a.$$

Therefore, $b=a$ and our definitions coincide for simple functions.

Proposition 1.3 *Let f and g be positive measurable functions.*

(a) $\int cf d\mu = c\int f d\mu$ for any real $c \geq 0$.
(b) If $0 \leq f \leq g$, then $\int f d\mu \leq \int g d\mu$.

Proof of Proposition 1.3
We first prove (a). If $c = 0$ there is nothing to prove. Assume now that $c > 0$.

$$\int cf d\mu = \sup\left\{\int s d\mu : s \in A(cf)\right\}.$$

Note now that s is in $A(cf)$ if and only if $\frac{s}{c}$ is in $A(f)$. Hence, $A(cf) = cA(f)$, where $cA(f)$ is defined as $\{cs : s \in A(f)\}$. Therefore,

$$\left\{\int s d\mu : s \in A(cf)\right\} = \left\{\int s d\mu : s \in cA(f)\right\} = \left\{\int cs d\mu : s \in A(f)\right\}.$$

By Proposition 1.1 (a) we have for a simple function s that

$$\int cs d\mu = c\int s d\mu.$$

Hence,

$$\int cf d\mu = \sup\left\{c\int s d\mu : s \in A(f)\right\}.$$

It is easy to check that if A is a subset of the reals, then

$$\sup\{cx : x \in A\} = c\sup\{x : x \in A\}.$$

provided $c \geq 0$. Hence,

$$\int cf d\mu = c\sup\left\{\int s d\mu : s \in A(f)\right\} = c\int f d\mu.$$

This proves (a).

We now turn to (b). Observe that if $f \leq g$, then $A(f) \subset A(g)$. Therefore,

$$\sup\left\{\int s d\mu : s \in A(f)\right\} \leq \sup\left\{\int s d\mu : s \in A(g)\right\}.$$

That is,

$$\int f d\mu \leq \int g d\mu.$$

This completes the proof of Proposition 1.3.

1.3 The Monotone Convergence Theorem and Fatou's Lemma

We have yet to show that $\int(f+g)d\mu = \int fd\mu + \int gd\mu$. To prove this we will use the following fundamental theorem.

Theorem 1.1 (Monotone Convergence Theorem) *Let (f_n) be a sequence of positive measurable functions. Assume that for every $n \geq 1$ we have $f_n \leq f_{n+1}$. Then*

$$\lim_{n\to\infty} \int f_n d\mu = \int \lim_{n\to\infty} f_n d\mu.$$

In words, under the assumptions of the Monotone Convergence Theorem we can interchange the limit and the integral.

Proof of Theorem 1.1
We first check that $\lim_{n\to\infty} f_n(x)$ exists for every x in X. For a fixed x, the sequence of real numbers $(f_n(x))$ is an increasing sequence. Hence, $\lim_{n\to\infty} f_n(x) = \sup_n f_n(x)$ exists (and is possibly $+\infty$). Let f be defined on X by

$$f(x) = \sup_n f_n(x) = \lim_{n\to\infty} f_n(x).$$

Note that f is a positive measurable function (why?).

Observe now that for every $n \geq 1$, $f_n \leq f$. Hence, by Proposition 1.3 (b)

$$\int f_n d\mu \leq \int f d\mu. \tag{1.2}$$

For every $n \geq 1$, $f_n \leq f_{n+1}$ and therefore $\int f_n d\mu \leq \int f_{n+1} d\mu$. Thus, the sequence of real numbers $\left(\int f_n d\mu\right)$ is increasing. Hence, $\lim_{n\to\infty} \int f_n d\mu$ exists and is equal to $\sup_n \int f_n d\mu$. Let n go to infinity in (1.2) to get

$$\lim_{n\to\infty} \int f_n d\mu \leq \int f d\mu = \int \lim_{n\to\infty} f_n d\mu. \tag{1.3}$$

We now prove the reverse inequality. Let $0 < a < 1$, s be a fixed simple function such that $0 \leq s \leq f$ and

$$A_n = \{x \in X : f_n(x) \geq as(x)\}.$$

Note that for every $n \geq 1$, A_n is measurable (why?). Let $x \in A_n$, then

$$f_{n+1}(x) \geq f_n(x) \geq as(x).$$

Therefore, x is in A_{n+1}. That is, the sequence (A_n) of measurable sets is increasing.

Note that

$$f_n \geq f_n \mathbf{1}_{A_n} \geq as\mathbf{1}_{A_n}.$$

By Proposition 1.3,

$$\int f_n d\mu \geq \int f_n \mathbf{1}_{A_n} d\mu \geq \int as\mathbf{1}_{A_n} d\mu. \tag{1.4}$$

Let ν be defined by

$$\nu(A) = \int s\mathbf{1}_A d\mu$$

for A in $\mathcal{M}$. By (1.4),

$$\int f_n d\mu \geq a\nu(A_n). \tag{1.5}$$

By Proposition 1.2, ν is a measure on $\mathcal{M}$. Since the sequence (A_n) is increasing we have by Proposition 1.2 in Chapter 11

$$\lim_{n\to\infty} \nu(A_n) = \nu\left(\bigcup_{n\geq 1} A_n\right).$$

We now show that $\bigcup_{n\geq 1} A_n = X$. There are two possibilities.

If $f(x) = \sup_n f_n(x) = 0$, then $s(x) = 0$ and x belongs to every A_n.

If $f(x) > 0$, then $as(x) < f(x)$. Since $f(x) = \sup_n f_n(x)$ there must be a natural n_0 such that

$$as(x) < f_{n_0}(x) \leq f(x).$$

Thus, x belongs to A_{n_0}.

In both cases x is in $\bigcup_{n\geq 1} A_n$. Therefore, $\bigcup_{n\geq 1} A_n = X$. Letting n go to infinity in (1.5),

$$\lim_{n\to\infty} \int f_n d\mu \geq a\nu(X) = a\int s d\mu.$$

We now let a approach 1 to get

$$\lim_{n\to\infty} \int f_n d\mu \geq \int s d\mu.$$

Since the last inequality holds for any simple function s in $A(f) = \{s \text{ simple } : 0 \leq s \leq f\}$, $\lim_{n\to\infty} \int f_n d\mu$ is an upper bound of the set $\left\{\int s d\mu : s \in A(f)\right\}$. Thus, $\lim_{n\to\infty} \int f_n d\mu$ is larger than the least upper bound of that set which is $\int f d\mu$. That is,

$$\lim_{n\to\infty} \int f_n d\mu \geq \int f d\mu.$$

This together with (1.3) completes the proof of Theorem 1.1.

The Monotone Convergence Theorem will be applied many times. We will use the abbreviation MCT for it. This theorem provides a sufficient condition to interchange limit and integral. We will see other theorems with different sufficient conditions. However, as the next example shows the interchange of limit and integral need not hold.

Example 1.2 Consider the Lebesgue measure space $(\mathbb{R}, \mathcal{L}, m)$. Let $f_n = \mathbf{1}_{(n,n+1)}$. Let x be a real. By the Archimedean Property there exists a natural number N such that $N > x$. Hence, for $n \geq N$ we have $f_n(x) = 0$. Therefore, the sequence $(f_n(x))$ converges to 0. On the other hand, $\int f_n dm = 1$ for every $n \geq 1$. Thus,

$$\lim_{n\to\infty} \int f_n dm = 1 \neq \int \lim_{n\to\infty} f_n dm = 0.$$

Our first application of the Monotone Convergence Theorem is the following,

Proposition 1.4 *Let f and g be positive measurable functions. Then,*

$$\int (f+g) d\mu = \int f d\mu + \int g d\mu.$$

Proof of Proposition 1.4
Recall from Theorem 2.1 in Chapter 10 that every measurable function is the pointwise limit of an increasing sequence of simple functions. Hence, there exist two sequences (s_n) and (t_n) of simple functions such that for every x the sequences $(s_n(x))$ and $(t_n(x))$ of real numbers are increasing and converge to $f(x)$ and $g(x)$, respectively.

By Proposition 1.1 (b) we have for every $n \geq 1$,

$$\int (s_n + t_n) d\mu = \int s_n d\mu + \int t_n d\mu. \tag{1.6}$$

The MCT can be applied to the three increasing sequences $(s_n + t_n)$, (s_n), and (t_n) to get

$$\lim_{n\to\infty}\int (s_n+t_n)d\mu = \int (f+g)d\mu,$$

$$\lim_{n\to\infty}\int s_n d\mu = \int f d\mu,$$

$$\lim_{n\to\infty}\int t_n d\mu = \int g d\mu.$$

We use these three limits in (1.6) to get

$$\int (f+g)d\mu = \int f d\mu + \int g d\mu.$$

This completes the proof of Proposition 1.6.

An interesting consequence of Proposition 1.4 and the MCT is the following

Corollary 1.1 *Let (f_n) be a sequence of positive measurable functions. Then,*

$$\int \left(\sum_{n=1}^{\infty} f_n\right) d\mu = \sum_{n=1}^{\infty}\int f_n d\mu.$$

Proof of Corollary 1.1
Let g_n be the partial sum

$$g_n = \sum_{k=1}^{n} f_k.$$

For every $n \geq 1$, $g_n \geq 0$ and $g_{n+1} \geq g_n$ (why?). Hence, for every x in X $\lim_n g_n(x)$ exists. Let f be defined on X by

$$f(x) = \lim_n g_n(x) = \sum_{k=1}^{\infty} f_k(x).$$

Note that the MCT applies to the sequence (g_n). Hence,

$$\lim_{n\to\infty}\int g_n d\mu = \int f d\mu.$$

Using Proposition 1.4 an easy induction proof on n shows that for every $n \geq 1$

$$\int g_n d\mu = \sum_{k=1}^{n}\int f_k d\mu.$$

Since the l.h.s. has a limit so does the r.h.s. Therefore,

$$\lim_{n\to\infty}\int g_n d\mu = \sum_{k=1}^{\infty}\int f_k d\mu.$$

Using that $\lim_{n\to\infty}\int g_n d\mu = \int f d\mu$ we have

$$\int f d\mu = \sum_{k=1}^{\infty}\int f_k d\mu.$$

The proof of Corollary 1 is complete.

We now turn to another fundamental result.

Theorem 1.2 (Fatou's Lemma) *Let (f_n) be a sequence of positive measurable functions. Then,*

$$\int \left(\liminf_{n\to\infty} f_n\right) d\mu \le \liminf_{n\to\infty}\int f_n d\mu.$$

Fatou's Lemma only requires the functions to be measurable and positive. But in contrast to the MCT it only provides an inequality.

Proof of Theorem 1.2
Recall that for any sequence (a_n) of real numbers we have that

$$\liminf_{n\to\infty} a_n = \lim_{n\to\infty} b_n,$$

where $b_n = \inf_{k\ge n} a_k$ for every $n \ge 1$. The sequence (b_n) is increasing. Similarly, for every x and $n \ge 1$ let

$$g_n(x) = \inf_{k\ge n} f_k(x).$$

Then,

$$\lim_{n\to\infty} g_n(x) = \liminf_{n\to\infty} f_n(x).$$

For all x in X and all $n \ge 1$ we have $g_n(x) \le f_n(x)$. Therefore,

$$\int g_n d\mu \le \int f_n d\mu.$$

Hence,

$$\liminf_{n\to\infty}\int g_n d\mu \le \liminf_{n\to\infty}\int f_n d\mu. \tag{1.7}$$

Since (g_n) is an increasing sequence of positive measurable functions (why?) the sequence $\left(\int g_n d\mu\right)$ has a limit. Hence,

$$\liminf_{n\to\infty}\int g_n d\mu = \lim_{n\to\infty}\int g_n d\mu.$$

On the other hand, the MCT applies to (g_n) and yields

$$\lim_{n\to\infty}\int g_n d\mu = \int \lim_{n\to\infty} g_n d\mu.$$

Using the last equality in (1.7)

$$\int \lim_{n\to\infty} g_n d\mu \le \liminf_{n\to\infty}\int f_n d\mu.$$

Since $\lim_{n\to\infty} g_n = \liminf_{n\to\infty} f_n$, the proof of Theorem 1.2 is complete.

1.4 Null Sets and Integration

The following inequality will turn out to be quite useful. It connects integral and measure.

Proposition 1.5 (Chebyshev's Inequality) *Let $f \ge 0$ be measurable. Then, for any $a \ge 0$ we have*

$$\int f d\mu \ge a\mu(f > a).$$

Recall that the notation $\{f > a\}$ is a convenient abbreviation for

$$\{x \in X : f(x) > a\}.$$

Proof of Proposition 1.5
Let $E = \{f > a\}$, then E is a measurable set. Since for every x, $a\mathbf{1}_E(x) \le f(x)$ (why?) we have,

$$\int a\mathbf{1}_E d\mu \le \int f d\mu.$$

Since

$$\int a\mathbf{1}_E d\mu = a\mu(E) = a\mu(f > a),$$

the proof of Proposition 1.5 is complete.

A property will be said to be true **almost everywhere** if the set where the property is not true is a null set. For instance we will say that $f = g$ a.e. if

$$\mu\left(\{x \in X : f(x) \neq g(x)\}\right) = 0.$$

A property may be true a.e. with respect to a measure μ and not true a.e. with respect to another measure. If any confusion is possible we will write μ a.e. instead of a.e.

Proposition 1.6 *Let f be positive and measurable. Then, $\int f d\mu = 0$ if and only if $f = 0$ a.e.*

Proof of Proposition 1.6
We first prove the direct implication. Assume that $\int f d\mu = 0$. We apply Chebyshev's inequality with $a = 1/n$ to get

$$\int f d\mu \geq \frac{1}{n}\mu(f > \frac{1}{n}).$$

For $n \geq 1$, let $A_n = \{f > \frac{1}{n}\}$. Since $\int f d\mu = 0$ we have by the inequality above that $\mu(A_n) = 0$ for every $n \geq 1$. Observe also that the sequence of measurable sets (A_n) is increasing. Hence,

$$\lim_{n\to\infty} \mu(A_n) = \mu\left(\bigcup_{n\geq 1} A_n\right) = 0.$$

Note that $\bigcup_{n\geq 1} A_n = \{f > 0\}$ (why?). Therefore, $\mu(f > 0) = 0$. Since we are assuming that $f \geq 0$ we must have $f = 0$ a.e. (why?). This proves the direct implication.

We now turn to the converse. Assume that $f = 0$ a.e. We first prove that the integral of f is 0 when f is simple. Let

$$f = \sum_{i=1}^{n} a_i \mathbf{1}_{A_i}$$

where $n \geq 1$, $a_i \geq 0$ is a real number and A_i is measurable for $1 \leq i \leq n$. If $a_i > 0$ for some i we must have $\mu(A_i) = 0$ (otherwise $f > 0$ on a non-null set contradicting $f = 0$ a.e.). If $a_i = 0$, then $a_i\mu(A_i) = 0$. Therefore, for all i we have $a_i\mu(A_i) = 0$. Hence,

$$\int f d\mu = \sum_{i=1}^{n} a_i \mu(A_i) = 0.$$

This proves the property for f simple. We now turn to the general case. Let $f \geq 0$ such that $f = 0$ a.e. Let s be a simple function such that $0 \leq s \leq f$. Then, $s = 0$ a.e. (why?) and by the special case we just proved $\int s d\mu = 0$. Hence, the set

$$\left\{ \int s d\mu : s \text{ simple and } 0 \leq s \leq f \right\}$$

is the singleton $\{0\}$. Since $\int f d\mu$ is defined as the sup of that set we get $\int f d\mu = 0$. This completes the proof of Proposition 1.6.

As the next result shows, if the integral is finite, then the function is finite a.e.

Proposition 1.7 *Assume $f \geq 0$ is measurable and $\int f d\mu < +\infty$. Then, $f(x) < +\infty$ for almost every x.*

Proof of Proposition 1.7
Let $A_n = \{f > n\}$ for every $n \geq 1$. By Chebyshev's inequality, for every $n \geq 1$

$$\mu(A_n) \leq \frac{1}{n} \int f d\mu. \tag{1.8}$$

Since the integral of f is finite,

$$\lim_{n \to \infty} \frac{1}{n} \int f d\mu = 0,$$

and therefore $\lim_{n \to \infty} \mu(A_n) = 0$.

Note now that (A_n) is a decreasing sequence of measurable sets (why?). Letting $n = 1$ in (1.8) we get that $\mu(A_1) < +\infty$. Therefore we can apply Proposition 1.2 in Chapter 11 to get

$$\lim_{n \to \infty} \mu(A_n) = \mu\left(\bigcap_{n \geq 1} A_n\right) = 0.$$

Note that

$$\bigcap_{n \geq 1} A_n = \{f = +\infty\}.$$

Hence, $\mu(f = +\infty) = 0$. That is, the set on which f takes infinite values is null. This completes the proof of Proposition 1.7.

Proposition 1.8 *Assume that $f \geq 0$ and $g \geq 0$ are measurable and that $f = g$ a.e. Then,*

$$\int f d\mu = \int g d\mu.$$

Proposition 1.8 tells us that functions can be modified on a null set without affecting the integral.

Proof of Proposition 1.8
Let $E = \{f = g\}$. By hypothesis $\mu(E^c) = 0$. We have

$$f = f\mathbf{1}_E + f\mathbf{1}_{E^c}.$$

By additivity of the integral we have

$$\int f d\mu = \int f\mathbf{1}_E d\mu + \int f\mathbf{1}_{E^c} d\mu.$$

Note that $\{f\mathbf{1}_{E^c} > 0\} \subset E^c$ which is a null set. Hence, $f\mathbf{1}_{E^c} = 0$ a.e. and by Proposition 1.6, $\int f\mathbf{1}_{E^c} d\mu = 0$. Therefore,

$$\int f d\mu = \int f\mathbf{1}_E d\mu = \int g\mathbf{1}_E d\mu.$$

Since f and g play symmetric roles we also have

$$\int g d\mu = \int g\mathbf{1}_E d\mu.$$

Hence, $\int f d\mu = \int g d\mu$. This completes the proof of Proposition 1.8.

Problems

1.

(a) Show that if $A \subset \mathbb{R}$ and $c \geq 0$, then

$$\sup\{cx : x \in A\} = c \sup\{x : x \in A\}.$$

(b) Show that if $A \subset B \subset \mathbb{R}$, then

$$\sup\{x : x \in A\} \leq \sup\{x : x \in B\}.$$

2. Consider the Lebesgue measure space $(\mathbb{R}, \mathcal{L}, m)$. For $n \geq 1$, let $f_n = n\mathbf{1}_{(0,1/n)}$.

(a) Show that (f_n) converges pointwise.
(b) Show that $\lim_n \int f_n dm$ exists.
(c) Show that

$$\lim_{n\to\infty} \int f_n dm \neq \int \lim_{n\to\infty} f_n dm.$$

3. Assume that (f_n) is a sequence of positive measurable functions. Show that for every $n \geq 1$

$$\int \left(\sum_{k=1}^{n} f_k \right) d\mu = \sum_{k=1}^{n} \int f_k d\mu.$$

4. Assume that (f_n) is a sequence of positive measurable functions. Show that

$$\lim_n \sum_{k=1}^{n} \int f_k d\mu$$

exists.

5. Assume that the sequence of positive measurable functions (f_n) converges to some f.

(a) Show that

$$\int f d\mu \leq \liminf_{n\to\infty} \int f_n d\mu.$$

(b) Give an example for which the inequality in (a) is strict.

6. Let f be a positive measurable function. For every natural $n \geq 1$ let $f_n = \min(f, n)$. Show that

$$\lim_{n\to\infty} \int f_n d\mu = \int f d\mu.$$

7. Assume that (f_n) is a decreasing sequence of positive measurable functions. That is, for every x in X and $n \geq 1$ we have $f_n(x) \geq f_{n+1}(x)$.

(a) Assume that $\int f_1 d\mu < +\infty$. Show that

$$\lim_{n\to\infty} \int f_n d\mu = \int \lim_{n\to\infty} f_n d\mu.$$

(Consider $g_n = f_1 - f_n$.)

(b) Give an example showing that the interchange of limit and integral need not hold if $\int f_1 d\mu = +\infty$.

8. Let f be a positive measurable function defined on the measure space $(X, \mathcal{M}, \mu)$. For A in $\mathcal{M}$, define $\nu(A) = \int_A f d\mu$. Show that ν is a measure on $\mathcal{M}$.

9. Let f be a measurable function and $E = \{f > a\}$. Show that $a\mathbf{1}_E \leq f$.

10. Let $0 \leq g \leq f$ be measurable functions. Show that if $f = 0$ a.e., then $g = 0$ a.e.

11. Let f be a measurable function. For $n \geq 1$, let $A_n = \{f > \frac{1}{n}\}$.

(a) Show that for every $n \geq 1$, A_n is measurable.
(b) Show that the sequence (A_n) is increasing.
(c) Show that

$$\bigcup_{n\geq 1} A_n = \{f > 0\}.$$

12. Let f be a measurable function. For $n \geq 1$, let $A_n = \{f > n\}$.

(a) Show that for every $n \geq 1$, A_n is measurable.
(b) Show that the sequence (A_n) is decreasing.
(c) Show that

$$\bigcap_{n\geq 1} A_n = \{f = +\infty\}.$$

13. Let f be a measurable function such that $\mu(f > 1) > 0$. Show that there exists $a > 1$ such that $\mu(f > a) > 0$.

2 The Integral of a Real Valued Function

2.1 The Definition of the Integral

We introduce a new notation. Let x be a real number we define

$$x^+ = \max(x, 0) \text{ and } x^- = \max(-x, 0).$$

Note that x^+ and x^- are both positive or 0 and that

$$x = x^+ - x^- \text{ and } |x| = x^+ + x^-.$$

Moreover, it is easy to see that

$$x^+ \leq |x| \text{ and } x^- \leq |x|.$$

Consider a measurable function $f : X \longrightarrow \mathbb{R}$. We let for every x in X

$$f^+(x) = (f(x))^+ \text{ and } f^-(x) = (f(x))^-.$$

The functions f^+ and f^- are both positive and measurable (why?). Moreover,

$$f = f^+ - f^- \text{ and } |f| = f^+ + f^-.$$

Definition 2.1 *Consider a measurable function $f : X \longrightarrow \mathbb{R}$. The function f is said to be integrable if $\int |f| d\mu < +\infty$. The set of measurable and integrable functions on $(X, \mathcal{M}, \mu)$ is denoted by $L^1(\mu)$. Moreover, if f is integrable we define the integral of f as*

$$\int f d\mu = \int f^+ d\mu - \int f^- d\mu.$$

Note that the three integrals $\int |f| d\mu$, $\int f^+ d\mu$, and $\int f^- d\mu$ are defined since the three corresponding functions are measurable and positive. The difference $\int f^+ d\mu - \int f^- d\mu$ on the other hand could be $+\infty - \infty$ which is undefined. We now show that both integrals in the difference are finite for integrable functions and therefore, our definition of $\int f d\mu$ is meaningful. Observe that

$$f^+ \leq |f| \text{ and } f^- \leq |f|.$$

Since all the functions involved are positive we have by Proposition 1.3

$$\int f^+ d\mu \leq \int |f| d\mu \text{ and } \int f^- d\mu \leq \int |f| d\mu.$$

Assuming that f is integrable, $\int |f| d\mu$ is finite and so are $\int f^+ d\mu$ and $\int f^- d\mu$.

Proposition 2.1 *Assume that f belongs to $L^1(\mu)$. Then there exists a function $\tilde{f}$ in $L^1(\mu)$ such that*

(a) $f = \tilde{f}$ a.e.
(b) $\tilde{f}$ is finite everywhere.
(c) $\int \tilde{f} d\mu = \int f d\mu$.

Proposition 2.1 allows us to replace an integrable function by a function which is finite everywhere without changing the integral of the original function. This will be helpful when we deal with the addition of two integrable functions f and g. For instance there may be some x such that $f(x) + g(x) = +\infty - \infty$ which is undefined. For integration purposes we can replace f and g by functions that are finite everywhere and therefore avoid this issue.

Proof of Proposition 2.1
If f is in $L^1(\mu)$, then by Proposition 1.7 $|f|$ (and therefore f) is finite except possibly on a null set E. We set $\tilde{f} = f\mathbf{1}_{E^c}$. Note that $\tilde{f}$ is measurable and finite everywhere (why?). Since E is a null set we have $f = \tilde{f}$ a.e.

By Proposition 1.8

$$\int (\tilde{f})^+ d\mu = \int f^+ d\mu \text{ and } \int (\tilde{f})^- d\mu = \int f^- d\mu.$$

Hence, $\int \tilde{f} d\mu = \int f d\mu$. This completes the proof of Proposition 2.1.

2.2 Properties of the Integral

We start with linearity.

Proposition 2.2 *The integral is linear. That is,*

(a) Let f be in $L^1(\mu)$ and c be a real number, then cf is in $L^1(\mu)$ and

$$\int cf d\mu = c \int f d\mu.$$

(b) Let f and g be in $L^1(\mu)$. Then, $f + g$ is in $L^1(\mu)$ and

$$\int (f + g) d\mu = \int f d\mu + \int g d\mu.$$

Proof of Proposition 2.2
We prove (a) first. We have by Proposition 1.3 that

$$\int |cf| d\mu = |c| \int |f| d\mu < +\infty.$$

Hence, cf is integrable if f is integrable.

Assume that $c \geq 0$. By definition,

$$\int cf d\mu = \int (cf)^+ d\mu - \int (cf)^- d\mu.$$

If $c \geq 0$ and x is a real, then $(cx)^+ = cx^+$ and $(cx)^- = cx^-$. Hence, for $c \geq 0$ we have

$$\int cf d\mu = \int cf^+ d\mu - \int cf^- d\mu.$$

Since $c \geq 0$, $f^+ \geq 0$, and $f^- \geq 0$, we may apply Proposition 1.4 to get

$$\int cf d\mu = c \int f^+ d\mu - c \int f^- d\mu = c \int f d\mu.$$

On the other hand, if $c < 0$, then $(cx)^+ = -cx^-$ and $(cx)^- = -cx^+$. Hence,

$$\int cf d\mu = \int (-cf^-) d\mu - \int (-cf^+) d\mu.$$

Using that $-c > 0$ we can again use Proposition 1.4 to get

$$\int cf d\mu = -c \int f^- d\mu + c \int f^+ d\mu = c \int f d\mu.$$

This completes the proof of part (a).

We now turn to part (b). By Proposition 2.1 f and g can be changed on a null set so that $f + g$ is almost everywhere equal to a function which is defined everywhere. We abuse the notation by denoting the new function by the same $f + g$.

Since

$$|f + g| \leq |f| + |g|,$$

we have by Propositions 1.3 and 1.4 that

$$\int |f + g| d\mu \leq \int (|f| + |g|) d\mu = \int |f| d\mu + \int |g| d\mu < +\infty.$$

Therefore, if f and g are integrable so is $f + g$.

It is not difficult to check that for all real numbers a and b we have

$$(a + b)^+ + a^- + b^- = (a + b)^- + a^+ + b^+.$$

Let f and g be two integrable functions. Then,

$$(f + g)^+ + f^- + g^- = (f + g)^- + f^+ + g^+.$$

Since all these functions are positive we can use Proposition 1.4 to get

$$\int (f + g)^+ d\mu + \int f^- d\mu + \int g^- d\mu = \int (f + g)^- d\mu + \int f^+ d\mu + \int g^+ d\mu.$$

The functions f, g, and $f + g$ are integrable so all the integrals above are finite. Hence,

$$\int (f+g)^+ d\mu - \int (f+g)^- d\mu = \int f^+ d\mu - \int f^- d\mu + \int g^+ d\mu - \int g^- d\mu.$$

That is,

$$\int (f+g) d\mu = \int f d\mu + \int g d\mu.$$

Proposition 2.3 *Let f be in $L^1(\mu)$. Then,*

$$\left| \int f d\mu \right| \le \int |f| d\mu.$$

Proof of Proposition 2.3
By the definition of the integral and the triangle inequality we have

$$\left| \int f d\mu \right| = \left| \int f^+ d\mu - \int f^- d\mu \right| \le \left| \int f^+ d\mu \right| + \left| \int f^- d\mu \right|.$$

Since f^+ and f^- are positive so are their integrals and therefore

$$\left| \int f d\mu \right| \le \int f^+ d\mu + \int f^- d\mu = \int (f^+ + f^-) d\mu.$$

Since $|f| = f^+ + f^-$ we have

$$\left| \int f d\mu \right| \le \int |f| d\mu.$$

The proof of Proposition 2.3 is complete.

Proposition 2.4 *Let f and g be in $L^1(\mu)$. Then, $\int_E f d\mu = \int_E g d\mu$ for every measurable set E if and only if $f = g$ a.e.*

Proof of Proposition 2.4
Assume that $\int_E f d\mu = \int_E g d\mu$ for every measurable set E. Let $E = \{f > g\}$, then E is measurable (why?). We have

$$\int_E (f-g) d\mu = \int \mathbf{1}_E (f-g) d\mu = 0.$$

Since $\mathbf{1}_E(f-g) \ge 0$, by Proposition 1.6 we have that $\mathbf{1}_E(f-g) = 0$ a.e. Observe that $\mathbf{1}_E(x)(f-g)(x) = 0$ if and only if $1_E(x) = 0$. Hence, $1_E = 0$ a.e. That is, $f \le g$ a.e. By interchanging f and g the same arguments show that $g \le f$ a.e. Therefore, $f = g$ a.e. This proves the direct implication.

We now turn to the converse. Let E be a measurable set. Then, by Proposition 2.3

$$\left|\int_E f d\mu - \int_E g d\mu\right| = \left|\int (f-g)\mathbf{1}_E d\mu\right| \le \int |f-g|\mathbf{1}_E d\mu \le \int |f-g| d\mu$$

where the last inequality is a consequence of $|f-g|\mathbf{1}_E \le |f-g|$. Since $|f-g| = 0$ a.e. we have by Proposition 1.6, $\int |f-g|d\mu = 0$. Hence,

$$\left|\int_E f d\mu - \int_E g d\mu\right| = 0.$$

This completes the proof of Proposition 2.4.

Proposition 2.5 *Let f and g be in $L^1(\mu)$. Assume that $f \le g$ a.e., then*

$$\int f d\mu \le \int g d\mu.$$

Proof of Proposition 2.5
Let $E = \{f \le g\}$. Then $(g-f)\mathbf{1}_E = g - f$ a.e. (why?). By Proposition 2.4 we have

$$\int_E (g-f) d\mu = \int (g-f) d\mu.$$

Since $(g-f)\mathbf{1}_E \ge 0$, $\int_E (g-f)d\mu \ge 0$. Hence,

$$\int (g-f) d\mu = \int g d\mu - \int f d\mu \ge 0.$$

This completes the proof of Proposition 2.5.

2.3 The Dominated Convergence Theorem

The next result is the third of the three fundamental theorems of the Lebesgue's theory. The two other results being the Monotone Convergence Theorem and Fatou's Lemma. Note that these three results are all concerned with interchange of integral and limit.

Theorem 2.1 (Dominated Convergence Theorem) *Let (f_n) be a sequence of functions in $L^1(\mu)$. Assume that the sequence (f_n) converges almost everywhere to some measurable function f. Assume also that there exists a g in $L^1(\mu)$ such that for all $n \ge 1$*

$$|f_n(x)| \leq g(x) \text{ almost everywhere.}$$

Then, f belongs to $L^1(\mu)$ and

$$\lim_{n\to\infty} \int f_n d\mu = \int \lim_{n\to\infty} f_n d\mu = \int f d\mu.$$

Proof of Theorem 2.1
We first prove the theorem in the particular case when the hypotheses hold everywhere. That is, for every x, $(f_n(x))$ converges to $f(x)$ and for all $n \geq 1$, $|f_n(x)| \leq g(x)$.

Since every f_n is measurable so is $\lim_n f_n$ (see Corollary 2.1 in Chapter 10).

By letting n go to infinity in

$$|f_n(x)| \leq g(x),$$

we get $|f(x)| \leq g(x)$. Therefore,

$$\int |f| d\mu \leq \int g d\mu < +\infty.$$

Hence, f is in $L^1(\mu)$.

Since $-g \leq f_n \leq g$ we have that $f_n + g \geq 0$. By Fatou's Lemma

$$\int \liminf_n (f_n + g) d\mu \leq \liminf_n \int (f_n + g) d\mu. \tag{2.1}$$

Recall that if c is a real and (a_n) is a sequence of real numbers, then

$$\liminf_n (c + a_n) = c + \liminf_n a_n.$$

Therefore, $\liminf_n (f_n + g) = f + g$ and

$$\liminf_n \int (f_n + g) d\mu = \liminf_n \left(\int f_n d\mu + \int g d\mu \right) = \liminf_n \int f_n d\mu + \int g d\mu.$$

Using this in (2.1)

$$\int (f + g) d\mu \leq \liminf_n \int f_n d\mu + \int g d\mu.$$

Since $\int g d\mu < +\infty$ we can subtract this term on both sides of the inequality above to get

$$\int f d\mu \leq \liminf_n \int f_n d\mu. \tag{2.2}$$

We now apply Fatou's Lemma to the sequence $(g - f_n)$ to get

$$\int \liminf_n (g - f_n) d\mu \leq \liminf_n \int (g - f_n) d\mu.$$

Therefore,

$$\int (g - f) d\mu \leq \int g d\mu + \liminf_n \left(- \int f_n d\mu \right).$$

Recall that if (a_n) is a real sequence, then

$$\liminf_n (-a_n) = -\limsup a_n.$$

Hence,

$$\int g d\mu - \int f d\mu \leq \int g d\mu - \limsup_n \int f_n d\mu.$$

Therefore,

$$\limsup_n \int f_n d\mu \leq \int f d\mu.$$

This together with (2.2) yield

$$\liminf_n \int f_n d\mu = \limsup_n \int f_n d\mu = \int f d\mu.$$

That is, $\lim_n \int f_n d\mu$ exists and is equal to $\int f d\mu$. This proves the Dominated Convergence Theorem in the particular case when the hypotheses hold everywhere.

We now turn to the general case. Let

$$E_1 = \{x \in X : (f_n(x)) \text{ converges to } f(x)\}$$

and

$$E_2 = \{x \in X : f_n(x) \leq g(x) \text{ for all } n \geq 1\}.$$

Since E_1^c and E_2^c are null sets so is $E_1^c \cup E_2^c$ (why?). Let $E = E_1 \cap E_2$. We have that $(f_n \mathbf{1}_E)$ converges everywhere to $f \mathbf{1}_E$ and that for all $n \geq 1$, $|f_n \mathbf{1}_E| \leq g 1_E$ everywhere. Hence, the particular case of the Dominated Convergence Theorem just proved applies. We get

$$\lim_{n \to \infty} \int_E f_n d\mu = \int_E \lim_{n \to \infty} f_n d\mu. \tag{2.3}$$

Since E^c is a null set we have for every $n \geq 1$

$$\int_E f_n d\mu = \int f_n d\mu,$$

and

$$\int_E \lim_n f_n d\mu = \int \lim_n f_n d\mu.$$

Note that there is a slight difficulty in the equality above. The sequence (f_n) need not converge on E^c. We can just set $\lim_n f_n = 0$ on E^c. Since this is a null set it does not affect the integral $\int \lim_n f_n d\mu$. Using the two last equalities in (2.3) yields,

$$\lim_{n\to\infty} \int f_n d\mu = \int \lim_{n\to\infty} f_n d\mu.$$

This completes the proof of Theorem 2.1.

There will be multiple applications of the Dominated Convergence Theorem. We will use the abbreviation DCT. Our first application is the following,

Corollary 2.1 *Let (f_n) be a sequence of functions in $L^1(\mu)$ such that*

$$\sum_{n\geq 1} \int |f_n| d\mu < +\infty.$$

Then, the series $\sum_{n\geq 1} f_n$ converges a.e. to a function in $L^1(\mu)$. Moreover,

$$\sum_{n\geq 1} \int f_n d\mu = \int \sum_{n\geq 1} f_n d\mu.$$

Proof of Corollary 2.1
Let $g = \sum_{n\geq 1} |f_n|$. Note that for any x, $g(x)$ is a positive terms infinite series. Hence, g is defined at any x but can take infinite values. Moreover, g is measurable (why?). By Corollary 1.1

$$\int g d\mu = \sum_{n\geq 1} \int |f_n| d\mu < +\infty.$$

Hence, g is in $L^1(\mu)$. By Proposition 1.7, g is finite a.e. That is, for almost all x the series $\sum_{n\geq 1} |f_n(x)|$ converges. Hence, for almost all x the series $\sum_{n\geq 1} f_n(x)$ converges. Let $g_n = \sum_{k=1}^n f_k$ and $f = \sum_{k\geq 1} f_k$, then (g_n) converges to f a.e. We have

$$|g_n| \leq \sum_{k=1}^{n} |f_k| \leq \sum_{k \geq 1} |f_k| = g.$$

Since g is in $L^1(\mu)$ the DCT applies. Hence, f is in $L^1(\mu)$ and

$$\lim_{n \to \infty} \int g_n d\mu = \int f d\mu = \int \left(\sum_{k \geq 1} f_k \right) d\mu. \tag{2.4}$$

Since for every $n \geq 1$ we have $\int g_n d\mu = \sum_{k=1}^{n} \int f_k d\mu$ (why?) we get

$$\lim_{n \to \infty} \int g_n d\mu = \sum_{k=1}^{\infty} \int f_k d\mu. \tag{2.5}$$

By (2.4) and (2.5),

$$\sum_{n \geq 1} \int f_n d\mu = \int \sum_{n \geq 1} f_n d\mu.$$

This completes the proof of Corollary 2.1.

Problems

Unless otherwise specified $(X, \mathcal{M}, \mu)$ is a complete measure space in the problems below.

1. Let x be a real number.

(a) Show that $x^+ \geq 0$ and $x^- \geq 0$.
(b) Show that $x = x^+ - x^-$.
(c) Show that $|x| = x^+ + x^-$.

2. Show that if $f = g$ a.e., then $f^+ = g^+$ a.e.
(Prove and use that $|a^+ - b^+| \leq |a - b|$ for all real numbers a and b).

3.

(a) Let a and b be positive real numbers. Show that $|a - b| = a + b$ if and only if $a = 0$ or $b = 0$.
(b) Let f be in $L^1(\mu)$. Use (a) to show that

$$\left| \int f d\mu \right| = \int |f| d\mu$$

if and only if $f = f^+$ a.e. or $f = f^-$ a.e.

4. Let f and g be measurable and such that $f \leq g$ a.e. and $g \leq f$ a.e. Show that $f = g$ a.e.

5. Assume that the measure space $(X, \mathcal{M}, \mu)$ is finite. That is, $\mu(X) < +\infty$.

(a) Assume that (f_n) is a sequence of measurable functions that converges uniformly to some f. That is, $\sup_x |f_n(x) - f(x)|$ converges to 0 as n goes to infinity. Show that

$$\lim_n \int f_n d\mu = \int f d\mu.$$

(b) Give an example that shows that the convergence in (a) need not hold if $\mu(X) = +\infty$.

6. Assume that $\mu(X) < +\infty$.

(a) Let (f_n) be a sequence of measurable functions. Assume that for almost every x, the sequence $(f_n(x))$ of real numbers converges to some $f(x)$ and that there exists M such that for every $n \geq 1$, $f_n(x) \leq M$ a.e. Show that

$$\lim_n \int f_n d\mu = \int f d\mu.$$

(b) Give an example that shows that the convergence in (a) need not hold if $\mu(X) = +\infty$.

7. Let (f_n) and (g_n) be sequences of measurable functions that converge a.e. to f and g, respectively. Assume that for every $n \geq 1$, g_n is integrable and so is g. Moreover, assume that for every $n \geq 1$ we have $|f_n| \leq g_n$ a.e. and that

$$\lim_n \int g_n d\mu = \int g d\mu.$$

(a) Show that for every $n \geq 1$, f_n is integrable and so is f.
(b) Prove that

$$\lim_n \int f_n d\mu = \int f d\mu.$$

(Follow the pattern used in the proof of the DCT).

8. Assume that the sequence (g_n) is in $L^1(\mu)$ and that (g_n) converges a.e. to g which is also in $L^1(\mu)$. Moreover, assume $\lim_n \int |g_n| d\mu = \int |g| d\mu$. Show that for every measurable set A we have

$$\lim_n \int_A g_n d\mu = \int_A g d\mu.$$

(Use Problem 7..)

9. Assume that the sequence (h_n) is in $L^1(\mu)$ and that (h_n) converges a.e. to h which is also in $L^1(\mu)$.

(a) Show that

$$\lim_n \int |h_n| d\mu = \int |h| d\mu \text{ if and only if } \lim_n \int |h_n - h| d\mu = 0.$$

(Use Problem 7.)

(b) Give an example showing that (a) need not be true if h is not integrable.

10. Assume that f is in $L^1(\mu)$. For every $n \geq 1$, let $f_n = \min(|f|, n)$. Let $\epsilon > 0$.

(a) Show that there exists an n such that

$$0 \leq \int (|f| - |f_n|) d\mu < \frac{\epsilon}{2}.$$

(b) Show the existence of a $\delta > 0$ such that if $\mu(A) < \delta$, then

$$\int_A |f| d\mu < \epsilon.$$

11. Assume that the sequence (f_n) and the function f are in $L^1(\mu)$. Assume also that for every $n \geq 1$

$$\int |f_n - f| d\mu \leq a_n$$

where the real series $\sum_{n \geq 1} a_n$ converges. Show that (f_n) converges to f a.e.

12. Assume that $\mu(X) = 1$ and that f is an integrable function with

$$\int f d\mu = 1.$$

Show that if $0 < a < 1$, then

$$\int_{\{f > a\}} f d\mu > 1 - a.$$

13. Assume that $\mu(X) < +\infty$, let f be a positive measurable function.

(a) Show that $\lim_n \int f^n d\mu$ exists.

(b) Show that

$$\lim_n \int f^n d\mu < +\infty$$

if and only if $\mu(f > 1) = 0$.

14. Assume that $\mu(X) < +\infty$, let (f_n) be a sequence of measurable functions that converges to f a.e.

(a) Let $\phi : \mathbb{R} \longrightarrow \mathbb{R}$ be a bounded continuous function. Show that

$$\lim_n \int \phi(f_n) d\mu = \int \phi(f) d\mu.$$

(b) Show that

$$\lim_n \int |f_n| \exp(-|f_n|)) d\mu = \int |f| \exp(-|f|)) d\mu.$$

15. Let f be an integrable function and (E_n) be a sequence of measurable disjoint sets. Show that

$$\sum_{k \geq 1} \int_{E_k} f d\mu = \int_E f d\mu,$$

where $E = \bigcup_{k \geq 1} E_k$.

Chapter 13
Integrals with Respect to Counting Measures

1 The Integral

Consider the set of natural numbers $\mathbb{N}$ equipped with the $\sigma-$algebra $\mathcal{P}(\mathbb{N})$ (i.e., all the subsets of $\mathbb{N}$). With this $\sigma-$algebra every subset of $\mathbb{N}$ is measurable and every function $a : \mathbb{N} \longrightarrow \mathbb{R}$ is measurable as well.

Let (w_n) be a sequence of positive real numbers. We define the counting measure ν_w with respect to w by

$$\nu_w(A) = \sum_{k \in A} w_k,$$

for any $A \subset \mathbb{N}$. We already checked that ν_w is indeed a measure, see Example 1.1 in Chapter 11.

Next we introduce the Lebesgue integral with respect to ν_w. It will be convenient to switch back and forth between the function and sequence notations. That is, we will think of $a : \mathbb{N} \longrightarrow \mathbb{R}$ as a function or as a sequence (a_n).

Proposition 1.1 *Let (a_n) be a positive sequence. Then,*

$$\int a d\nu_w = \sum_{k=1}^{\infty} w_k a_k.$$

That is, the Lebesgue integral of a is simply a series! Note that a positive terms series is always defined but may be infinite.

Proof of Proposition 1.1
Let $a : \mathbb{N} \longrightarrow \mathbb{R}$ be a positive function. Let (s_n) be a sequence of functions defined as follows. For $n \geq 1$ we have

$$s_n(k) = a_k \text{ for } 1 \leq k \leq n,$$

R. B. Schinazi, *From Classical to Modern Analysis*,
https://doi.org/10.1007/978-3-319-94583-5_13

and $s_n(k) = 0$ for $k > n$. We have

$$s_n = \sum_{k=1}^{n} a_k \mathbf{1}_{\{k\}}.$$

This shows that for every $n \geq 1$, s_n is a simple function. By the definition of the integral of a simple function (Chapter 12, Section 1.1),

$$\int s_n d\nu_w = \sum_{k=1}^{n} a_k \nu_w(k) = \sum_{k=1}^{n} a_k w_k.$$

Hence,

$$\lim_n \int s_n d\nu_w = \sum_{k=1}^{+\infty} a_k w_k,$$

where the limit is possibly infinite. Observe now that (s_n) is an increasing (why?) and positive sequence. Observe also that for every $k \geq 1$, $(s_n(k))$ converges to a_k (why?). Hence, the Monotone Convergence Theorem applies and

$$\lim_n \int s_n d\nu_w = \int \lim_n s_n d\nu_w = \int a d\nu_w.$$

Therefore,

$$\int a d\nu_w = \lim_n \int s_n d\nu_w = \sum_{k=1}^{+\infty} w_k a_k.$$

This completes the proof of Proposition 1.1.

- By definition a function a is integrable with respect to the counting measure ν_w if and only if $\int |a| d\nu_w < +\infty$. By Proposition 1.1, a is integrable if and only if

$$\int |a| d\nu_w = \sum_{k=1}^{+\infty} w_k |a_k| < +\infty.$$

2 Interchanging the Order of Summation

In this section we will give sufficient conditions to interchange the order of summation for infinite sums. Our first result shows that we can always interchange the order for positive term double sequences.

Proposition 2.1 *Assume that $a_n(k) \geq 0$ for all $n \geq 1$ and all $k \geq 1$. Then,*

$$\sum_{k\geq 1}\sum_{n\geq 1} a_n(k) = \sum_{n\geq 1}\sum_{k\geq 1} a_n(k).$$

Proof of Proposition 2.1
Let $w_k = 1$ for all $k \geq 1$. Let ν be the counting measure defined by

$$\nu(A) = \sum_{k\in A} w_k$$

for A in $\mathcal{P}(\mathbb{N})$. For $k \geq 1$, let

$$b(k) = \sum_{n\geq 1} a_n(k).$$

In particular, b is a positive function with possibly infinite values. By Corollary 1.1 in Chapter 12 and the definition of b,

$$\sum_{n\geq 1}\int a_n d\nu = \int \left(\sum_{n\geq 1} a_n\right) d\nu = \int b d\nu. \tag{2.1}$$

By Proposition 1.1 and the definitions of ν and b,

$$\int b d\nu = \sum_{k\geq 1} b(k) = \sum_{k\geq 1}\sum_{n\geq 1} a_n(k). \tag{2.2}$$

Since

$$\int a_n d\nu = \sum_{k\geq 1} a_n(k),$$

we have

$$\sum_{n\geq 1}\int a_n d\nu = \sum_{n\geq 1}\sum_{k\geq 1} a_n(k). \tag{2.3}$$

Using (2.2) and (2.3) in (2.1) yields

$$\sum_{n\geq 1}\sum_{k\geq 1} a_n(k) = \sum_{k\geq 1}\sum_{n\geq 1} a_n(k).$$

This completes the proof of Proposition 2.1.

Proposition 2.1 applies to positive double sequences. In the general case we have,

Proposition 2.2 *Assume that*

$$\sum_{n\geq 1}\sum_{k\geq 1}|a_n(k)| < +\infty.$$

Then,

$$\sum_{k\geq 1}\sum_{n\geq 1}a_n(k) = \sum_{n\geq 1}\sum_{k\geq 1}a_n(k),$$

where both sides are finite.

Because we are dealing with terms which are not necessarily positive there are convergence questions involved in this result. Proposition 2.2 ensures that all the series (and the series of series) converge!

Proof of Proposition 2.2
Recall that a consequence of the DCT is that if (a_n) is a sequence in $L^1(\nu)$ such that

$$\sum_{n\geq 1}\int |a_n|d\nu < +\infty,$$

then the series $\sum_{n\geq 1} a_n$ converges a.e. to an integrable function and

$$\sum_{n\geq 1}\int a_n d\nu = \int \sum_{n\geq 1} a_n d\nu, \tag{2.4}$$

see Corollary 2.1 in Chapter 12. This result applies here since

$$\sum_{n\geq 1}\int |a_n|d\nu = \sum_{n\geq 1}\sum_{k\geq 1}|a_n(k)|,$$

which is finite by assumption. By Proposition 1.1 and (2.4) we get

$$\sum_{n\geq 1}\sum_{k\geq 1}a_n(k) = \sum_{k\geq 1}\sum_{n\geq 1}a_n(k).$$

This completes the proof of Proposition 2.2.

We now give an example showing that interchanging summation order need not be true.

Example 2.1 Consider the double sequence $(a_n(k))$ for $k \geq 1$ and $n \geq 1$.

$$a_n(k) = 1 \text{ for } n = k$$
$$a_n(k) = -1 \text{ for } n = k+1$$
$$a_n(k) = 0 \text{ otherwise.}$$

Note that for $k \geq 1$ we have

$$\sum_{n\geq 1} a_n(k) = a_1(k) + a_2(k) + \ldots a_k(k) + a_{k+1}(k) + \cdots = 0.$$

Since this is true for every $k \geq 1$ we get

$$\sum_{k\geq 1}\sum_{n\geq 1} a_n(k) = 0.$$

On the other hand, for $n = 1$ we have

$$\sum_{k\geq 1} a_1(k) = a_1(1) + a_1(2) + \cdots = 1.$$

For $n \geq 2$ we get

$$\sum_{k\geq 1} a_n(k) = a_n(1) + a_n(2) + \cdots + a_{n-1}(n) + a_n(n) + \cdots = 0.$$

Therefore,

$$\sum_{n\geq 1}\sum_{k\geq 1} a_n(k) = 1.$$

In particular,

$$\sum_{k\geq 1}\sum_{n\geq 1} a_n(k) \neq \sum_{n\geq 1}\sum_{k\geq 1} a_n(k).$$

Problems

1. For every $k \geq 0$, $(a_n(k))$ is a sequence. Apply the Monotone Convergence Theorem to give sufficient conditions on $(a_n(k))$ so that

$$\lim_n \sum_{k\geq 0} a_n(k) = \sum_{k\geq 0} \lim_n a_n(k).$$

2. Apply the Dominated Convergence Theorem to give sufficient conditions on $(a_n(k))$ so that

$$\lim_n \sum_{k\geq 0} a_n(k) = \sum_{k\geq 0} \lim_n a_n(k).$$

3. Let $w_k = 1$ for all $k \geq 1$. Consider the measure space $(\mathbb{N}, \mathcal{P}(\mathbb{N}, \nu)$ where ν is the counting measure with respect to (w_k).

(a) Show that every function $a : \mathbb{N} \longrightarrow \mathbb{R}$ is measurable in this measure space.
(b) Show that the only null set is the empty set.
(c) Show that if (a_n) converges a.e. to a, then it converges everywhere.

4. Consider Example 2.1. Show that this example does not contradict Propositions 2.1 and 2.2.

5. Assume that for every $k \geq 0$ and $n \geq 1$, $a_n(k) \geq 0$. Assume that for every $k \geq 0$ the sequence $(a_n(k))$ converges to some ℓ_k. Show that

$$\sum_{k\geq 0} \ell_k \leq \liminf_n \sum_{k\geq 0} a_n(k).$$

Chapter 14
Riemann and Lebesgue Integrals

In this chapter our measure space will be the Lebesgue measure space $(\mathbb{R}, \mathcal{L}, m)$, where $\mathcal{L}$ is the Lebesgue σ-algebra and m the Lebesgue measure. We will say that a function f is Lebesgue integrable on some measurable set I if $\int_I |f| dm < +\infty$.

We now introduce some notation regarding the Riemann integral.

1 The Riemann Integral

A partition $\mathcal{P}$ of the interval $I = [a, b]$ is a finite sequence of real numbers

$$\mathcal{P} = \{x_0 = a < x_1 < \cdots < x_n = b\}.$$

The mesh of $\mathcal{P}$ is defined by $\max\{x_1 - x_0, x_2 - x_1, \ldots, x_n - x_{n-1}\}$.

Let f be a bounded function on $I = [a, b]$. We define the **lower and upper Darboux sums** of f on $\mathcal{P}$ by,

$$s(f, \mathcal{P}) = \sum_{k=1}^{n} m_i(x_{i-1} - x_i)$$

$$S(f, \mathcal{P}) = \sum_{k=1}^{n} M_i(x_{i-1} - x_i).$$

where $m_i = \inf\{f(x) : x \in [x_{i-1}, x_i]\}$ and $M_i = \sup\{f(x) : x \in [x_{i-1}, x_i]\}$.

The function f is said to be **Riemann integrable** on $[a, b]$ if it is bounded and has the following property. Let $(\mathcal{P}_n)$ be a sequence of partitions of $[a, b]$ such that the mesh of $\mathcal{P}_n$ converges to 0 as n goes to infinity. Then,

R. B. Schinazi, *From Classical to Modern Analysis*,
https://doi.org/10.1007/978-3-319-94583-5_14

$$\lim_n s(f, \mathcal{P}_n) = \lim_n S(f, \mathcal{P}_n).$$

Moreover, the limit above defines the Riemann integral of f on $[a, b]$.

2 Comparing the Riemann and Lebesgue Integrals

Our first result shows that Lebesgue integration generalizes Riemann integration.

Theorem 2.1 Let f be a bounded function on $I = [a, b]$. If f is Riemann integrable on I, then f is Lebesgue integrable on I and the two integrals coincide.

The indicator function $f = \mathbf{1}_{\mathbb{Q}}$ of the rational numbers is measurable and almost everywhere 0 on [0, 1] with respect to the Lebesgue measure. Hence, f is Lebesgue integrable and its integral is 0. On the other hand, it is easy to see that for any partition $\mathcal{P}$ we have $s(f, \mathcal{P}) = 0$ and $S(f, \mathcal{P}) = 1$. Hence, f is not Riemann integrable. This shows that the converse of Theorem 2.1 does not hold. In fact, the Lebesgue integral generalizes the Riemann integral to a (much) wider class of functions.

Proof of Theorem 2.1
For a function f and a partition $\mathcal{P}$ we define a lower and an upper step functions by

$$L(f, \mathcal{P}) = \sum_{k=1}^{n} m_i \mathbf{1}_{[x_{i-1}, x_i]}$$

$$U(f, \mathcal{P}) = \sum_{k=1}^{n} M_i \mathbf{1}_{[x_{i-1}, x_i]}.$$

Since f is bounded all m_i and M_i are finite. Observe that for any partition $\mathcal{P}$, the step functions $L(f, \mathcal{P})$ and $U(f, \mathcal{P})$ are Lebesgue measurable (why?) and are bounded on I. Hence, they are Lebesgue integrable on I. Using the linearity of the integral we compute the following Lebesgue integrals (recall that the Lebesgue measure of an interval is its length),

$$\int L(f, \mathcal{P}) dm = s(f, \mathcal{P}) \quad \text{and} \quad \int U(f, \mathcal{P}) dm = S(f, \mathcal{P}),$$

where $s(f, \mathcal{P})$ and $S(f, \mathcal{P})$ are lower and upper Darboux sums we defined above.

For $n \geq 1$, let $\mathcal{P}_n$ be a partition of $[a, b]$. We assume that for all $n \geq 1$, $\mathcal{P}_n \subset \mathcal{P}_{n+1}$ and that the mesh of $\mathcal{P}_n$ converges to 0 as n goes to infinity. It is easy to see that for every x in $[a, b]$ the sequence $(L(f, \mathcal{P}_n)(x))$ is increasing and the sequence $(U(f, \mathcal{P}_n)(x))$ is decreasing. These two sequences are also bounded. Therefore, they both converge. We denote the limits by $g(x)$ and $h(x)$, respectively. As limits of measurable functions, g and h are measurable. Since f is bounded there exists M such that for every $n \geq 1$ and x in I

$$|L(f, \mathcal{P}_n)(x)| \leq M.$$

A constant M is Lebesgue integrable on I. Hence, the DCT applies and

$$\lim_n s(f, \mathcal{P}_n) = \lim_n \int_I L(f, \mathcal{P}_n) dm = \int_I g dm.$$

Similarly, we have

$$\lim_n S(f, \mathcal{P}_n) = \lim_n \int_I U(f, \mathcal{P}_n) dm = \int_I h dm.$$

Since f is Riemann integrable we have that

$$\lim_n s(f, \mathcal{P}_n) = \lim_n S(f, \mathcal{P}_n) = \int_a^b f(x) dx,$$

where the last integral is the Riemann integral. Therefore,

$$\int_I g dm = \int_I h dm = \int_a^b f(x) dx. \tag{2.1}$$

Using that $h - g \geq 0$ on I and $\int (h - g) dm = 0$ we have that $h - g = 0$ a.e. on I.

Since, for every $n \geq 1$ and x in I

$$L(f, \mathcal{P}_n)(x) \leq f(x) \leq U(f, \mathcal{P}_n)(x).$$

Letting n go to infinity,

$$g(x) \leq f(x) \leq h(x).$$

Observe that

$$\{f \neq h\} = \{f < h\} \subset \{h \neq g\}.$$

Since $h = g$ a.e. the set $\{h \neq g\}$ is null. Let $A = \{f \neq h\}$. A is a subset of a null set. But the Lebesgue measure m is complete. Hence, A is Lebesgue measurable and is a null set with respect to the Lebesgue measure. That is, $f = h$ a.e.

We now prove that f is Lebesgue measurable. Let c be a real number. We have

$$\{f > c\} = (\{f > c\} \cap A) \cup \left(\{f > c\} \cap A^c\right).$$

Now, $\{f > c\} \cap A$ is a subset of A which is a null set. Hence, $\{f > c\} \cap A$ is Lebesgue measurable. Observe that $\{f > c\} \cap A^c = \{h > c\} \cap A^c$ which is Lebesgue measurable. Therefore, $\{f > c\}$ is Lebesgue measurable as a union of Lebesgue measurable sets. This proves that f is Lebesgue measurable.

Since f is a measurable bounded function, f is Lebesgue integrable on I. Using $h - f = 0$ a.e. yields $\int_I (h - f)dm = 0$. By (2.1),

$$\int_I f dm = \int_I h dm = \int_a^b f(x)dx.$$

This proves that the Lebesgue integral of f is equal to its Riemann integral. The proof of Theorem 2.1 is complete.

The following is a very nice characterization of Riemann integrability.

Theorem 2.2 Let f be a bounded function on $I = [a, b]$. Then, f is Riemann integrable on I if and only if f is continuous a.e. on I.

It is well known that a continuous function is Riemann integrable. Theorem 2.2 tells us that the **only** Riemann integrable functions are functions that are continuous except possibly on a null set. In contrast, the function $\mathbf{1}_{\mathbb{Q}}$ which is nowhere continuous is Lebesgue integrable!

Proof of Theorem 2.2
We first prove the direct implication. Assume that f is Riemann integrable on I. We use the notation and some of the results in the proof of Theorem 2.1. Let $(\mathcal{P}_n)$ be a partition of I, assume that $\mathcal{P}_n \subset \mathcal{P}_{n+1}$ for every $n \geq 1$, and that the mesh of $\mathcal{P}_n$ converges to 0 as n goes to infinity. We have shown in the proof of Theorem 2.1 that the sequences of step functions $(L(f, \mathcal{P}_n))$ and $(U(f, \mathcal{P}_n))$ converge pointwise on I to g and h, respectively. These functions are Lebesgue measurable and for all x in I

$$g(x) \leq f(x) \leq h(x).$$

We also proved that $f = g = h$ a.e. In particular, $N_1 = \{g < h\}$ is a null set.
Let

$$N_2 = \bigcup_{n \geq 1} \mathcal{P}_n.$$

Since each $\mathcal{P}_n$ is a finite collection of points the Lebesgue measure of N_2 is 0. That is, N_2 is a null set and so is $N = N_1 \cup N_2$. We will now show that f is continuous at all points outside N.

By contradiction assume that there exists a point y_0 in I but not in N such that f is not continuous at y_0. Then, there is an $\epsilon > 0$ such that for any $\delta > 0$ there exists y in I such that $|y - y_0| < \delta$ and $|f(y) - f(y_0)| \geq \epsilon$. For any $n \geq 1$, there exists two consecutive points $x_{i,n}$ and $x_{i+1,n}$ in $\mathcal{P}_n$ such that

$$x_{i,n} < y_0 < x_{i+1,n}.$$

Using that y_0 is not in $\mathcal{P}_n$, we can pick $\delta > 0$ so that

$$(y_0 - \delta, y_0 + \delta) \subset I_n = [x_{i,n}, x_{i+1,n}].$$

Hence, there exists y in I_n such that $|f(y) - f(y_0)| \geq \epsilon$. This implies that the difference between $\inf_{I_n} f$ and $\sup_{I_n} f$ must be at least ϵ. Therefore, for every $n \geq 1$

$$U(f, \mathcal{P}_n)(y_0) - L(f, \mathcal{P}_n)(y_0) \geq \epsilon.$$

Letting n go to infinity we get

$$h(y_0) - g(y_0) \geq \epsilon.$$

Hence, y_0 belongs to N_1. This is a contradiction since y_0 does not belong to N_1. Thus, f is continuous at y_0. This proves that f is continuous everywhere except possibly on N.

We now turn to the converse. Assume that f is continuous a.e. on I. Assume that $(\mathcal{P}_n)$ is an arbitrary sequence of partitions of $[a, b]$ such that the mesh of $\mathcal{P}_n$ converges to 0 as n goes to infinity. Assume that f is continuous at y_0. Let $n \geq 1$, y_0 belongs to $I_n = [x_{i,n}, x_{i+1,n}]$ where $x_{i,n}$ and $x_{i+1,n}$ are two consecutive points in $\mathcal{P}_n$. Let $\epsilon > 0$, $\inf_{I_n} f + \epsilon$ is not lower bound of f on I_n. Hence, there exists a y_n in I_n such that

$$f(y_n) - \epsilon < \inf_{I_n} f \leq f(y_0). \tag{2.2}$$

Since the mesh of $\mathcal{P}_n$ converges to 0 we have that (y_n) converges to y_0. Given that f is continuous at y_0, the sequence $(f(y_n))$ converges to $f(y_0)$. By (2.2), the sequence $(\inf_{I_n} f)$ converges to $f(y_0)$. This shows that the sequence $(L(f, \mathcal{P}_n)(y_0))$ converges to $f(y_0)$. This is true at all continuity points y_0 of f. Therefore, the sequence of lower step functions $(L(f, \mathcal{P}_n))$ converges a.e. to f.

The sequence $(L(f, \mathcal{P}_n))$ is bounded by a constant. Since a constant is Lebesgue integrable on I the DCT applies and we have

$$\lim_n \int_I L(f, \mathcal{P}_n) dm = \int_I f dm.$$

Similar arguments show the same convergence for upper step functions and we get

$$\lim_n \int_I U(f, \mathcal{P}_n) dm = \int_I f dm.$$

As noted in the proof of Theorem 2.1, $\int_I L(f, \mathcal{P}_n) dm$ and $\int_I U(f, \mathcal{P}_n) dm$ are the lower and upper Darboux sums, respectively. The fact that

$$\lim_n \int_I L(f, \mathcal{P}_n)dm = \lim_n \int_I U(f, \mathcal{P}_n)dm$$

implies that f is Riemann integrable. This completes the proof of Theorem 2.2.

Problems

For these problems, the measure space is the Lebesgue measure space $(\mathbb{R}, \mathcal{L}, m)$

1. Assume that f continuous a.e. on $[a, b]$. Let $(\mathcal{P}_n)$ is an arbitrary sequence of partitions of $[a, b]$ such that the mesh of $\mathcal{P}_n$ converges to 0 as n goes to infinity. Prove that

$$\lim_n \int U(f, \mathcal{P}_n)dm = \int f dm.$$

2. Let f be a positive continuous function on $[0, +\infty)$. Show that

$$\lim_n \int_0^n f(x)dx = \int_{[0,+\infty)} f dm,$$

where on the l.h.s. we have the Riemann integral and on the r.h.s. we have the Lebesgue integral.

3. Given a real a and a function f the translate of f by a is denoted by $\tau_a f$ and is defined by

$$\tau_a f(x) = f(x + a).$$

(a) Let ϕ be a simple function. Show that

$$\int \tau_a \phi dm = \int \phi dm.$$

(Recall that the Lebesgue measure m is translation invariant).

(b) Let $f \geq 0$ be a positive measurable function. Show that

$$\int \tau_a f dm = \int f dm.$$

(Use (a) and the fact that there is an increasing sequence of simple functions converging to f.)

(c) Show that if f is Lebesgue integrable then

$$\int \tau_a f dm = \int f dm.$$

4. Assume that f is Lebesgue integrable on [0, 1]. Compute

$$\lim_n \int_{[0,1]} x^n f(x) dm.$$

5. Compute

$$\lim_n \int_{[0,+\infty)} \exp(-2x)(1 + \frac{x}{n})^n dm.$$

6. Consider for $n \geq 1$ and $x \geq 0$, $f_n(x) = \exp(-nx) - 2\exp(-2nx)$.

(a) Show that

$$\sum_{n\geq 1} \int_{[0,+\infty)} f_n dm \neq \int_{[0,+\infty)} \sum_{n\geq 1} f_n dm.$$

(b) Use Corollary 2.1 in Chapter 12 to show that

$$\sum_{n\geq 1} \int_{[0,+\infty)} |f_n| dm = +\infty.$$

(c) Show (b) by a direct computation.

7. Assume that f is Lebesgue integrable on $[a, b]$. Let

$$g(x) = \int_{[a,x]} f dm.$$

(a) Show that g is defined on $[a, b]$.
(b) Show that g is continuous on (a, b). (Assume that (x_n) converges to c in (a, b) and show that $f\mathbf{1}_{[a,x_n]}$ converges a.e. to $f\mathbf{1}_{[a,c]}$.)

Chapter 15
Modes of Convergence

1 The Lebesgue Integral Metric

Let $(X, \mathcal{M}, \mu)$ be a measure space. Consider $L^1(\mu)$ the space of integrable functions and let

$$d(f, g) = \int |f - g| d\mu.$$

For all f, g, and h in $L^1(\mu)$ it is easy to check that $d(f, g) \geq 0$, $d(f, g) = d(g, f)$, and

$$d(f, g) \leq d(f, h) + d(h, g).$$

We also have that $d(f, f) = 0$. However, if $d(f, g) = 0$, it is not true that $f = g$! We only have that $f = g$ almost everywhere. Hence, d is not a metric on $L^1(\mu)$. We will still call d a metric but we will have to remember that an equality of two functions for this metric is an almost everywhere equality. There is actually a way to make d a genuine metric by replacing $L^1(\mu)$ by a different space, see the problems.

1.1 The Lebesgue Integral Metric Is Complete

One of the reasons the Riemann theory is not satisfactory is that it does not provide a complete metric, see Theorem 3.4 in Chapter 7. In contrast, we now show that the Lebesgue integral metric space is complete.

Theorem 1.1 Let $(X, \mathcal{M}, \mu)$ be a measure space. Let

R. B. Schinazi, *From Classical to Modern Analysis*,
https://doi.org/10.1007/978-3-319-94583-5_15

$$d(f,g) = \int |f - g| d\mu.$$

Then, $(L^1(\mu), d)$ is a complete metric space.

Proof of Theorem 1.1
Let (f_n) a Cauchy sequence in $(L^1(\mu), d)$. We have to show that there exists an f in $L^1(\mu)$ such that (f_n) converges to f.

Using the definition of a Cauchy sequence it is not difficult to see that there exists a strictly increasing sequence of natural numbers (n_i) such that if $n \geq n_i$ and $p \geq n_i$, then

$$d(f_n, f_p) < \frac{1}{2^i}.$$

For $i \geq 1$, let $g_i = f_{n_i}$. We have

$$d(g_i, g_{i+1}) < \frac{1}{2^i}. \tag{1.1}$$

Define

$$g = \sum_{i=1}^{\infty} |g_{i+1} - g_i|.$$

Note that g is measurable as a limit of measurable functions. We have

$$\int g d\mu = \int \left(\sum_{i=1}^{\infty} |g_{i+1} - g_i| \right) d\mu = \sum_{i=1}^{\infty} \int |g_{i+1} - g_i| d\mu,$$

where the interchange of sum and integral comes from Corollary 1.1 in Chapter 12. By (1.1) we have

$$\int g d\mu = \sum_{i=1}^{\infty} \int |g_{i+1} - g_i| d\mu = \sum_{i=1}^{\infty} d(g_i, g_{i+1}) < \sum_{i=1}^{\infty} \frac{1}{2^i} = 1.$$

Hence, g is integrable and therefore a.e. finite. Therefore, the series $\sum_{i=1}^{\infty}(g_{i+1} - g_i)$ converges absolutely a.e. Thus, the partial sum

$$\sum_{i=1}^{k} (g_{i+1} - g_i) = g_{k+1} - g_1$$

converges a.e. as k goes to infinity. Since $g_i = f_{n_i}$, this yields that the subsequence (f_{n_i}) of (f_n) converges a.e. to some measurable function f.

Now that we have a potential limit f, we need to show two properties.

- The whole sequence (f_n) converges to f with respect to the metric d.
- f is in $L^1(\mu)$.

Let $\epsilon > 0$, there exists a natural N such that if $n \geq N$ and $p \geq N$, then $d(f_n, f_p) < \epsilon$. Let $i \geq N$, then $n_i \geq N$ (why?) and therefore

$$d(f_{n_i}, f_p) = \int |f_{n_i} - f_p| d\mu < \epsilon,$$

for $p \geq N$. By Fatou's Lemma we have

$$\int \liminf_i |f_{n_i} - f_p| d\mu \leq \liminf_i \int |f_{n_i} - f_p| d\mu \leq \epsilon.$$

Using that $\liminf_i |f_{n_i} - f_p| = |f - f_p|$ a.e. (why?) we have for $p \geq N$

$$\int |f - f_p| d\mu \leq \epsilon. \tag{1.2}$$

That is, if $p \geq N$, we have $d(f, f_p) \leq \epsilon$. This shows that the whole sequence (f_n) converges to f with respect to the metric d.

We now show that f is in $L^1(\mu)$. By (1.2), $(f - f_p)$ belongs to $L^1(\mu)$. By hypothesis, f_p belongs to $L^1(\mu)$. Since

$$|f| \leq |f - f_p| + |f_p|,$$

f belongs to $L^1(\mu)$.

In summary, we have shown that if (f_n) is a Cauchy sequence in $(L^1(\mu), d)$, then there exists a function f in $L^1(\mu)$ such that (f_n) converges to f in $(L^1(\mu), d)$. This shows that $(L^1(\mu), d)$ is a complete metric space. The proof of Theorem 1.1 is complete.

The next result links convergence in $L^1(\mu)$ and a.e. convergence.

Theorem 1.2 Assume that the sequence (f_n) converges to some f in $(L^1(\mu), d)$. Then, there exists a subsequence (f_{n_i}) that converges to f a.e.

In words, convergence of a sequence in $(L^1(\mu), d)$ implies a.e. convergence of a subsequence. As we will see below the whole sequence need not converge a.e.

Proof of Theorem 1.2
If (f_n) converges in $(L^1(\mu), d)$ it must be a Cauchy sequence in that metric space. In the proof of Theorem 1.1 we showed that a Cauchy sequence has a subsequence (f_{n_i}) that converges a.e. to some integrable function h. We also proved that the whole sequence (f_n) converges to h in $(L^1(\mu), d)$. By assumption, (f_n) converges

to f in $(L^1(\mu), d)$. Since a limit is unique in a metric space we have $f = h$ a.e. Hence, (f_{n_i}) converges a.e. to f. This concludes the proof of Theorem 1.2.

The following two examples show that Theorem 1.2 is all we can say about the relation between L^1 convergence and a.e. convergence.

Example 1.1 Consider the Lebesgue measure m on the real line. For $n \geq 1$ let

$$f_n = \frac{1}{n}\mathbf{1}_{(0,n)}.$$

Show that (f_n) converges a.e. but not in $(L^1(m), d)$.

For any fixed $x \leq 0$ it is clear that $f_n(x) = 0$ for every $n \geq 1$ and therefore $(f_n(x))$ converges to 0. For $x > 0$ there exists $N > x$ (why?) and for $n \geq N$, $f_n(x) = \frac{1}{n}$. Hence, $(f_n(x))$ converges to 0. This shows that (f_n) converges to the null function everywhere.

By contradiction, assume that (f_n) converges to some f in $(L^1(m), d)$, where m is the Lebesgue measure on the real line. By Theorem 1.2 there exists a subsequence (f_{n_i}) that converges to f a.e. Since (f_{n_i}) converges to 0 everywhere we must have that $f = 0$ a.e. Thus, (f_n) converges to 0 in $(L^1(m), d)$. But for every $n \geq 1$,

$$d(f_n, 0) = \int |f_n - 0| dm = \frac{1}{n} m\left((0, n)\right) = \frac{1}{n} n = 1.$$

Hence, (f_n) does not converge to 0 in $(L^1(m), d)$. We have a contradiction. Therefore, (f_n) does not converge in $(L^1(\mu), d)$. Note that a very similar argument shows that no subsequence of (f_n) converges in $(L^1(\mu), d)$.

Example 1.2 Let (f_n) be defined by

$$f_1 = \mathbf{1}_{[0,1]}$$

$$f_2 = \mathbf{1}_{[0,1/2]} \quad f_3 = \mathbf{1}_{[1/2,1]}$$

$$f_4 = \mathbf{1}_{[0,1/4]} \quad f_5 = \mathbf{1}_{[1/4,1/2]} \quad f_6 = \mathbf{1}_{[1/2,3/4]} \quad f_7 = \mathbf{1}_{[3/4,1]}$$

and so on. More formally, for a fixed $n \geq 1$ there is a $k(n) \geq 0$ such $2^{k(n)} \leq n < 2^{k(n)+1}$. Hence, $n - 2^{k(n)} = j$ where $0 \leq j < 2^{k(n)}$. We define

$$f_n = \mathbf{1}_{[j/2^{k(n)}, (j+1)/2^{k(n)}]}.$$

Let m be the Lebesgue measure on [0, 1]. We have

$$\int |f_n| dm = m\left([j/2^{k(n)}, (j+1)/2^{k(n)}]\right) = 2^{-k(n)}.$$

As n goes to infinity so does $k(n)$ and therefore (f_n) converges to 0 in $(L^1(m), d)$.

We now show that (f_n) does not converge a.e. Let x be fixed in $[0, 1]$. There are infinitely many n such that $f_n(x) = 1$ and infinitely many n such that $f_n(x) = 0$ (why?). Hence, $(f_n(x))$ does not converge for any x in $[0, 1]$. That is, (f_n) converges nowhere!. However, we know by Theorem 1.2 that there is a subsequence of (f_n) that converges a.e.

1.2 The Completion of the Riemann Integral Metric

We denote by $\int_a^b f(x)dx$ the Riemann integral of a function f on $[a, b]$. Let $\mathcal{C}$ be the space of continuous functions on $[a, b]$. Let d_r be defined by

$$d_r(f, g) = \int_a^b |f(x) - g(x)|dx,$$

where f and g belong to $\mathcal{C}$. We know that $(\mathcal{C}, d_r)$ is not a complete space. That is, a Cauchy sequence in $(\mathcal{C}, d_r)$ need not converge in this space (see Theorem 3.4 in Chapter 7).

Consider the measure space $([a, b], \mathcal{L}, m)$ where $\mathcal{L}$ is the Lebesgue $\sigma-$ algebra and m is the Lebesgue measure on $[a, b]$. Let

$$d_m(f, g) = \int_{[a,b]} |f(x) - g(x)|dm,$$

for f and g in $L^1(m)$.

Theorem 1.3 The completion of $(\mathcal{C}, d_r)$ is $(L^1(m), d_m)$.

In words, the smallest complete metric space containing $(\mathcal{C}, d_r)$ is $(L^1(m), d_m)$. Theorem 1.3 shows that the Lebesgue integral is the right generalization of the Riemann integral!

Outline of the proof of Theorem 1.3
Let (f_n) be a Cauchy sequence in $(\mathcal{C}, d_r)$. It is also a Cauchy sequence in $(L^1(m), d_m)$. This is so because the Riemann and Lebesgue integrals coincide on continuous functions. By Theorem 1.1, (f_n) converges in $(L^1(m), d_m)$. That is, the limit exists and is in $L^1(m)$ but need not be in $\mathcal{C}$. This shows that the completion of $(\mathcal{C}, d_r)$ (that is, the smallest complete space containing $(\mathcal{C}, d_r)$) is in $(L^1(m), d_m)$. To complete the proof we need to show that the completion of $(\mathcal{C}, d_r)$ is exactly $(L^1(m), d_m)$. This is a consequence of the fact that $\mathcal{C}$ is dense in $(L^1(m), d_m)$. For a proof of this density result, see for instance Theorem 3.14 in Rudin (1987). This completes our (partial) proof of Theorem 1.3.

Problems

In all the problems below $(X, \mathcal{M}, \mu)$ is a measure space and d is the metric on $L^1(\mu)$ defined by

$$d(f, g) = \int |f - g| d\mu.$$

1. Let f and g be in $L^1(\mu)$ and a and b be two real numbers. Show that $af + bg$ is in $L^1(\mu)$.

2. Consider the sequence (f_n) from example 1.1. Show that no subsequence of (f_n) converges in $(L^1(\mu), d)$.

3. Assume that a sequence (f_n) converges to f a.e. and to g a.e. Show that $f = g$ a.e.

4. Assume that (f_n) converges to f in $(L^1(\mu), d)$ and that (f_n) converges to g a.e. Show that $f = g$ a.e.

5. Assume that (f_n) converges to f a.e. and that there exists an integrable function g such that $|f_n| \le g$ for all $n \ge 1$. Show that (f_n) converges in $(L^1(\mu), d)$.

6. Consider Example 1.2. Find an explicit subsequence of (f_n) that converges a.e.

7. Let $f \sim g$ if $f = g$ a.e.

(a) Check that this is an equivalence relation. That is, it is reflexive ($f \sim f$), it is symmetric (if $f \sim g$, then $g \sim f$) and it is transitive (if $f \sim g$ and $g \sim h$, then $f \sim h$).

(b) Show that if $f \sim f_1$ and $g \sim g_1$, then

$$d(f, g) = d(f_1, g_1).$$

Let $[f]$ be the equivalence class of a function f in $L^1(\mu)$. That is,

$$[f] = \{g \in L^1(\mu) : g \sim f\}.$$

Let $\mathcal{L}^1(\mu)$ be the space of all the equivalence classes. Define d on $\mathcal{L}^1(\mu)$ by

$$d([f], [g]) = d(f, g).$$

(c) Show that d is a metric on $\mathcal{L}^1(\mu)$.

2 Convergence in Measure

Let $(X, \mathcal{M}, \mu)$ be a measure space. A sequence (f_n) is said to converge **in measure** to some f if for every $\epsilon > 0$ we have

$$\lim_n \mu(|f_n - f| > \epsilon) = 0.$$

We first show that convergence in measure limits are unique up to a null set.

Proposition 2.1 *Consider a sequence of real-valued measurable functions. Assume that (f_n) converges in measure to f and to g, where f and g are measurable functions. Then, $f = g$ a.e.*

Proof of Proposition 2.1
Let $n \geq 1$ and $\epsilon > 0$. By the triangle inequality

$$\left(\{|f - f_n| \leq \frac{\epsilon}{2}\}\bigcap\{|f_n - g| \leq \frac{\epsilon}{2}\}\right) \subset \{|f - g| \leq \epsilon\}.$$

Taking the complements yields

$$\{|f - g| > \epsilon\} \subset \left(\{|f - f_n| > \frac{\epsilon}{2}\}\bigcup\{|f_n - g| > \frac{\epsilon}{2}\}\right).$$

Hence,

$$\mu(|f - g| > \epsilon) \leq \mu(|f - f_n| > \frac{\epsilon}{2}) + \mu(|f_n - g| > \frac{\epsilon}{2}).$$

Since (f_n) converges in measure to f and to g, both terms on r.h.s. converge to 0. We let n go to infinity in the preceding inequality to get

$$\mu(|f - g| > \epsilon) = 0,$$

for every $\epsilon > 0$. For $k \geq 1$, let

$$A_k = \{|f - g| > \frac{1}{k}\}.$$

The sets A_k are measurable (why?), the sequence (A_k) is increasing (why?) and $\mu(A_k) = 0$ for every $k \geq 1$ (why?). Hence,

$$\lim_k \mu(A_k) = \mu(\bigcup_{k\geq 1} A_k) = 0.$$

Since

$$\bigcup_{k\geq 1} A_k = \{f \neq g\},$$

we have that $f = g$ a.e. This completes the proof of Proposition 2.1.

Next we find connections between the three convergence notions of this chapter.

Proposition 2.2 *Assume that* (f_n) *converges to* f *in* $L^1(\mu)$. *Then,* (f_n) *converges in measure to* f.

Proof of Proposition 2.2
For $n \geq 1$ and $\epsilon > 0$, let $A = \{|f_n - f| > \epsilon\}$. Note that

$$|f_n - f| \geq \mathbf{1}_A |f_n - f| \geq \epsilon \mathbf{1}_A.$$

Integrating across the inequality yields

$$\int |f_n - f| d\mu \geq \epsilon \mu(A).$$

That is,

$$\mu(|f_n - f| > \epsilon) \leq \frac{1}{\epsilon} \int |f_n - f| d\mu.$$

Since the r.h.s. converges to 0 as n goes to infinity we have

$$\lim_n \mu(|f_n - f| > \epsilon) = 0.$$

Therefore, (f_n) converges in measure to f. This completes the proof of Proposition 2.2.

As the next example shows the converse of Proposition 2.2 is not true.

Example 2.1 Consider the Lebesgue measure m on the real line. For $n \geq 1$ let

$$f_n = n\mathbf{1}_{[0,1/n]}.$$

We will show that (f_n) converges in measure but not in $L^1(m)$.

Let $\epsilon > 0$ and $n \geq 1$, note that if $|f_n(x) - 0| > \epsilon$, then x belongs to $[0, 1/n]$. Hence,

$$m(|f_n - 0| > \epsilon) \leq m([0, 1/n]) = \frac{1}{n}.$$

Letting n go to infinity in the preceding equality shows that (f_n) converges in measure to 0.

We now turn to convergence in $L^1(m)$. By contradiction, assume that (f_n) converges to some f in $L^1(m)$. By Proposition 2.2, (f_n) converges in measure to f. Since we already know that (f_n) converges in measure to 0, by Proposition 2.1 we must have $f = 0$ a.e. That is, (f_n) converges to 0 in $L^1(m)$. But

$$\int |f_n - 0| dm = \int n \mathbf{1}_{[0,1/n]} dm = nm([0, 1/n]) = 1.$$

Therefore, (f_n) does not converge to 0 in $L^1(m)$! We have a contradiction. The sequence (f_n) does not converge in $L^1(m)$.

Proposition 2.3 *Let $(X, \mathcal{M}, \mu)$ be a finite measure space. That is, $\mu(X) < +\infty$. Assume that (f_n) converges to f a.e. Then, (f_n) converges in measure to f.*

In words, in a finite measure space almost everywhere convergence implies convergence in measure. We will give an example below showing that this need not be true if $\mu(X)$ is infinite.

Proof of Proposition 2.3
Let $\epsilon > 0$ and define for every $n \geq 1$, $A_n = \{|f_n - f| > \epsilon\}$. Then,

$$\mu(|f_n - f| > \epsilon) = \int \mathbf{1}_{A_n} d\mu.$$

For almost every x, $(f_n(x))$ converges to $f(x)$. Hence, for almost every x there exists N such that if $n \geq N$, then $|f_n(x) - f(x)| < \epsilon$. Thus, $\mathbf{1}_{A_n}(x) = 0$ for $n \geq N$. This shows that the sequence $(\mathbf{1}_{A_n})$ converges to 0 a.e. Moreover, for all $n \geq 1$

$$\mathbf{1}_{A_n} \leq 1.$$

Note now that 1 is integrable since $\mu(X) < +\infty$. Therefore, the DCT applies and we have

$$\lim_n \mu(|f_n - f| > \epsilon) = \lim_n \int \mathbf{1}_{A_n} d\mu = \int \lim_n \mathbf{1}_{A_n} d\mu = 0.$$

That is, (f_n) converges in measure to f. The proof of Proposition 2.3 is complete.

While the converse of Proposition 2.3 is not true, the following weaker result is.

Proposition 2.4 *Assume that (f_n) converges in measure to f. Then, there exists a subsequence (f_{n_i}) that converges a.e. to f.*

Proof of Proposition 2.4
Since (f_n) converges in measure to f it is not difficult to show the existence of natural numbers $n_1 < n_2 < \ldots$ such that for every natural i we have

$$\mu(|f_{n_i} - f| > \frac{1}{2^i}) < \frac{1}{2^i}.$$

For $i \geq 1$, let

$$B_i = \{|f_{n_i} - f| > \frac{1}{2^i}\}.$$

Observe that

$$\sum_{i \geq 1} \mu(B_i) < +\infty.$$

Therefore, we can apply the Borel-Cantelli Lemma (Theorem 1.1 in Chapter 11) to get

$$\mu(\limsup_i B_i) = 0.$$

Recall that

$$\limsup_i B_i = \bigcap_{j \geq 1} \bigcup_{i \geq j} B_i.$$

Since this is a null set almost every x belongs to the complement of $\limsup_i B_i$. Observe that

$$\left(\limsup_i B_i\right)^c = \bigcup_{j \geq 1} \bigcap_{i \geq j} B_i^c.$$

So, for almost every x there exists a $j \geq 1$ such that for all $i \geq j$, x belongs to B_i^c. That is, for $i \geq j$

$$|f_{n_i}(x) - f(x)| \leq \frac{1}{2^i}.$$

Letting i go to infinity yields

$$\lim_i f_{n_i}(x) = f(x).$$

This shows that (f_{n_i}) converges a.e. to f and completes the proof of Proposition 2.4.

Example 2.2 For $n \geq 1$, let

$$f_n = \mathbf{1}_{(n,n+1)}.$$

We will show that (f_n) converges a.e. but not in measure in the Lebesgue measure space. Note that $m(\mathbb{R}) = +\infty$ and therefore Proposition 2.3 does not apply.

Let x be a fixed real number. There exists a natural N such that $N > x$. Therefore, for all $n \geq N$, $f_n(x) = 0$. Hence, (f_n) converges everywhere to 0 on $\mathbb{R}$.

By contradiction assume that (f_n) converges in measure to some f. By Proposition 2.4 there exists a subsequence (f_{n_i}) that converges to f a.e. Since the whole sequence (f_n) converges to 0 we must have that $f = 0$ a.e. Hence, if (f_n) converges in measure it must converge in measure to 0.

Let $0 < \epsilon < 1$. Note that $f_n(x) > \epsilon$ if and only if x belongs to $(n, n+1)$. Hence,

$$m(|f_n - 0| > \epsilon) = m((n, n+1)) = 1.$$

Therefore, (f_n) does not converge in measure to 0. We have a contradiction. Hence, (f_n) does not converge in measure.

As the next result shows, under additional hypotheses convergence in measure implies convergence in $L^1(\mu)$.

Theorem 2.1 (Bounded Convergence Theorem) Assume that $\mu(X) < +\infty$ and let (f_n) be a sequence of measurable functions that converges in measure to some measurable function f. Moreover, assume that there exists a real $b > 0$ such that

$$|f_n| \leq b \text{ a.e.}$$

for all $n \geq 1$. Then, the sequence (f_n) converges to f in $L^1(\mu)$. That is,

$$\lim_n \int |f_n - f| d\mu = 0.$$

Proof of Theorem 2.1
We first show that $|f| \leq b + 1$ a.e. By the triangle inequality

$$\{|f| > b + 1\} \subset (\{|f - f_n| > 1\} \cup \{|f_n| > b\}),$$

for every $n \geq 1$. Therefore,

$$\mu(|f| > b + 1) \leq \mu(|f - f_n| > 1) + \mu(|f_n| > b).$$

The assumption $|f_n| \leq b$ a.e. is equivalent to $\mu(|f_n| > b) = 0$. Using that (f_n) converges in measure to f,

$$\lim_n \mu(|f - f_n| > 1) = 0.$$

Therefore,

$$\mu(|f| > b + 1) = 0.$$

Hence, $|f| \leq b+1$ a.e. With a little more work we can actually show that $|f| \leq b$ a.e. See the problems.

We now show that the sequence (f_n) converges in $L^1(\mu)$. Let $\epsilon > 0$ and

$$\alpha = \frac{\epsilon}{\mu(X)}.$$

Since $\mu(X) < +\infty$, α is strictly positive. We have

$$\int |f_n - f| d\mu = \int_{\{|f_n - f| \leq \alpha\}} |f_n - f| d\mu + \int_{\{|f_n - f| > \alpha\}} |f_n - f| d\mu.$$

Note that

$$\int_{\{|f_n - f| \leq \alpha\}} |f_n - f| d\mu \leq \alpha \mu(|f_n - f| \leq \alpha) \leq \alpha \mu(X) = \epsilon.$$

Since $|f| < b+1$ a.e. and $|f_n| < b$ a.e. we have

$$\int_{\{|f_n - f| > \alpha\}} |f_n - f| d\mu \leq (2b+1)\mu(|f_n - f| > \alpha).$$

Hence,

$$\int |f_n - f| d\mu \leq \epsilon + (2b+1)\mu(|f_n - f| > \alpha).$$

Using that (f_n) converges in measure to f we get

$$\limsup_n \int |f_n - f| d\mu \leq \epsilon.$$

Since this is true for every $\epsilon > 0$ we must have $\limsup_n \int |f_n - f| d\mu = 0$ and therefore

$$\lim_n \int |f_n - f| d\mu = 0.$$

The proof of Theorem 2.1 is complete.

Problems

In all the problems below $(X, \mathcal{M}, \mu)$ is a measure space and d is the metric on $L^1(\mu)$ defined by

$$d(f, g) = \int |f - g| d\mu.$$

1. Assume that (f_n) is a sequence of measurable functions converging to some f a.e. Assume that (f_n) has a subsequence (f_{n_k}) that converges to some g a.e. Show that $f = g$ a.e.

2. Let f and g be two measurable functions. For every $k \geq 1$ let

$$A_k = \{|f - g| > \frac{1}{k}\}.$$

(a) Show that A_k is measurable for every $k \geq 1$.
(b) Show that (A_k) is an increasing sequence of sets.
(c) Show that

$$\bigcup_{k \geq 1} A_k = \{f \neq g\}.$$

3. In this problem we prove **Egoroff's Theorem:** Assume $\mu(X) < +\infty$. Let (f_n) be a sequence of measurable functions converging a.e. to a measurable function f. Then, for any $\alpha > 0$ there exists a measurable set A such that (f_n) converges uniformly on A and $\mu(A^c) < \alpha$.

Let $\epsilon > 0$ and define for $n \geq 1$

$$E_n = \bigcup_{j \geq n} \{|f_j - f| > \epsilon\}.$$

(a) Show that

$$\lim_n \mu(E_n) = \mu(\bigcap_{n \geq 1} E_n).$$

(b) Show that $\lim_n \mu(E_n) = 0$.

For $k \geq 1$, let $\epsilon = \frac{1}{k}$ and denote the corresponding E_n by $E_n(k)$.

(c) Show the existence of natural numbers $n_1 < n_2 < \ldots$ such that

$$\mu(E_{n_k}) < \frac{\alpha}{2^k}.$$

Let

$$A^c = \bigcup_{k \geq 1} E_{n_k}.$$

(d) Show that $\mu(A^c) < \alpha$.
(e) Show that (f_n) converges uniformly on A.

4. The convergence in Egoroff's Theorem (see problem 3) is called **almost uniform convergence** Show that if a sequence is almost uniformly convergent, then it converges a.e. (Let D be the set of x's for which the sequence does converge. Show that $\mu(D) < \alpha$ for any $\alpha > 0$.)

5. Show that (f_n) converges to f in measure if and only if every subsequence of (f_n) converges in measure to f.

6. Assume that $\mu(X) < +\infty$ and let (f_n) be a sequence of measurable functions that converges in measure to some measurable function f. Moreover, assume that there exists a real $b > 0$ such that $|f_n| \leq b$ a.e. for all $n \geq 1$.

For $j \geq 1$, let

$$A_j = \{|f| > b + \frac{1}{j}\}.$$

(a) Show that for all natural numbers j and n we have

$$\mu(A_j) \leq \mu(|f - f_n| > \frac{1}{j}) + \mu(|f_n| > b) = \mu(|f - f_n| > \frac{1}{j}).$$

(b) Show that for every $j \geq 1$, $\mu(A_j) = 0$.
(c) Show that

$$\bigcup_{j \geq 1} A_j = \{|f| > b\}.$$

(d) Show that $\mu(|f| > b) = 0$.

7. Assume that the sequence (f_n) converges in measure to f. Assume that the function ϕ is such that

$$|\phi(x) - \phi(y)| \leq |x - y|$$

for all x and y. Show that the sequence $(\phi(f_n))$ converges in measure to $\phi(f)$.

8. Suppose that for every $\epsilon > 0$

$$\sum_{n \geq 1} \mu(|f_n - f| > \epsilon) < +\infty.$$

Show that (f_n) converges to f a.e.

9. Assume that (f_n) converges in measure to f. Show that every subsequence of (f_n) contains a further subsequence that converges to f a.e.

10. Let $\mu(X) < +\infty$. Assume that every subsequence of (f_n) contains a further subsequence that converges to f a.e. Show that (f_n) converges in measure to f. (Do a proof by contradiction).

11. Let $\mu(X) < +\infty$. Assume that (f_n) converges in measure to f. Let $g : \mathbb{R} \longrightarrow \mathbb{R}$ be a continuous function. Show that $(g \circ f_n)$ converges in measure to $g \circ f$. (Use Problems 9 and 10).

3 Uniform Integrability

A sequence (f_n) of measurable functions is said to be **uniformly integrable** if

$$\lim_{k\to+\infty} \sup_n \int_{\{|f_n|>k\}} |f_n| d\mu = 0.$$

Proposition 3.1 *Assume that $\mu(X) < +\infty$. If the sequence (f_n) is uniformly integrable, then for every $n \geq 1$, f_n is integrable and*

$$\sup_n \int |f_n| d\mu < +\infty.$$

Proof of Proposition 3.1
Since

$$\lim_{k\to+\infty} \sup_n \int_{\{|f_n|>k\}} |f_n| d\mu = 0$$

there exists a $k_0 > 0$ such that

$$\sup_n \int_{\{|f_n|>k_0\}} |f_n| d\mu < 1.$$

For any $n \geq 1$,

$$\begin{aligned}\int |f_n| d\mu &= \int_{\{|f_n|\leq k_0\}} |f_n| d\mu + \int_{\{|f_n|>k_0\}} |f_n| d\mu \\ &\leq k_0\mu(X) + 1.\end{aligned}$$

Since $\mu(X) < +\infty$, every f_n is integrable. The upper bound $k_0\mu(X) + 1$ is uniform in n, hence

$$\sup_n \int |f_n| d\mu \leq k_0 \mu(X) + 1 < +\infty.$$

This completes the proof of Proposition 3.1.

As the next example shows the converse of Proposition 3.1 does not hold.

Example 3.1 Consider the Lebesgue measure space on $X = [0, 1]$. For $n \geq 1$, let $f_n = n\mathbf{1}_{[0,1/n]}$. Then, for every $n \geq 1$, $\int |f_n| dm = 1$ and therefore $\sup_n \int |f_n| dm = 1$.

We now show that the sequence (f_n) is not uniformly integrable. Let $k > 0$, for $n > k$,

$$|f_n| = \mathbf{1}_{\{|f_n|>k\}} |f_n|.$$

Hence, for $n > k$ we have

$$\int_{\{|f_n|>k\}} |f_n| dm = \int |f_n| dm = 1.$$

In particular, $\sup_n \int_{\{|f_n|>k\}} |f_n| dm = 1$ for any $k > 0$. Thus, the sequence (f_n) is not uniformly integrable.

Proposition 3.2 *Let $p > 1$ and let (f_n) be a sequence of measurable functions. Assume that*

$$\sup_n \int |f_n|^p d\mu < +\infty.$$

Then, the sequence (f_n) is uniformly integrable.

Example 3.1 shows that $p = 1$ in Proposition 3.2 is not enough to yield uniform integrability.

Proof of Proposition 3.2
Let $k > 0$ and $n \geq 1$, we have

$$\begin{aligned} \int |f_n|^p d\mu &= \int_{\{|f_n|\leq k\}} |f_n|^p d\mu + \int_{\{|f_n|>k\}} |f_n|^p d\mu \\ &\geq \int_{\{|f_n|>k\}} |f_n|^p d\mu. \end{aligned}$$

Since $p > 1$, if $|f_n(x)| > k$, then

$$|f_n(x)|^p = |f_n(x)||f_n(x)|^{p-1} \geq |f_n(x)| k^{p-1}.$$

Therefore,

$$\int |f_n|^p d\mu \geq \int_{\{|f_n|>k\}} |f_n|^p d\mu \geq k^{p-1} \int_{\{|f_n|>k\}} |f_n| d\mu.$$

That is,

$$\int_{\{|f_n|>k\}} |f_n| d\mu \leq \frac{1}{k^{p-1}} \int |f_n|^p d\mu \leq \frac{1}{k^{p-1}} \sup_n \int |f_n|^p d\mu.$$

Therefore,

$$\sup_n \int_{\{|f_n|>k\}} |f_n| d\mu \leq \frac{1}{k^{p-1}} \sup_n \int |f_n|^p d\mu.$$

Since $\sup_n \int |f_n|^p d\mu < +\infty$ and $p > 1$ the r.h.s. goes to 0 as k goes to infinity. Hence,

$$\lim_{k\to+\infty} \sup_n \int_{\{|f_n|>k\}} |f_n| d\mu = 0.$$

The sequence (f_n) is uniformly integrable. The proof of Proposition 3.2 is complete.

We now state and prove two lemmas that will be helpful for the sequel and which are interesting in their own right.

Lemma 3.1 *Let f be integrable. Then,*

$$\lim_{k\to+\infty} \int_{\{|f|>k\}} |f| d\mu = 0.$$

Lemma 3.1 explains the name *uniform* integrability. The lemma shows that every integrable function f has the property $\lim_{k\to+\infty} \int_{\{f>k\}} |f| d\mu = 0$. That is, for every $\epsilon > 0$ there exists a k_0 such that if $k \geq k_0$, then

$$0 \leq \int_{\{f>k\}} |f| d\mu < \epsilon.$$

On the other hand, a sequence (f_n) is uniformly integrable when for every $\epsilon > 0$ there exists a k_0 such that if $k \geq k_0$

$$0 \leq \int_{\{|f_n|>k\}} |f_n| d\mu < \epsilon$$

for every $n \geq 1$. That is, k_0 is the *same* for every n. This is why the sequence is said to be *uniformly* integrable.

Proof of Lemma 3.1
For $n \geq 1$, let

$$g_n = f\mathbf{1}_{\{|f|>n\}}.$$

Since f is in $L^1(\mu)$, f is finite a.e. Hence, for almost every x there exists a natural n_x such that $|f(x)| < n_x$. In particular, $g_n(x) = 0$ for $n \geq n_x$. That is, for almost every x, the sequence $(g_n(x))$ converges to 0. Hence, the sequence (g_n) converges to 0 a.e.

Observe that for all $n \geq 1$, $|g_n| \leq |f|$. Therefore, the DCT applies and

$$\lim_n \int g_n d\mu = \int \lim_n g_n d\mu = 0.$$

That is,

$$\lim_n \int_{\{|f|>n\}} |f| d\mu = 0.$$

This proves the limit along natural numbers. For the general result, let $\epsilon > 0$ and let N be a natural such that if $n \geq N$, then

$$0 \leq \int_{\{|f|>n\}} |f| d\mu < \epsilon.$$

For $k > N$ we have

$$0 \leq \int_{\{|f|>k\}} |f| d\mu \leq \int_{\{|f|>N\}} |f| d\mu < \epsilon.$$

This proves that $\lim_{k\to+\infty} \int_{\{|f|>k\}} |f| d\mu = 0$. The proof of Lemma 3.1 is complete.

Lemma 3.2 *Let f be integrable. Then, for every $\epsilon > 0$ there exists a $\delta > 0$ such that if A is a measurable set and $\mu(A) < \delta$, then*

$$\int_A |f| d\mu < \epsilon.$$

Proof of Lemma 3.2
For $n \geq 1$, let $f_n = \min(|f|, n)$. Since f is integrable, for almost every x, $|f(x)|$ is finite. Hence, there exists a natural n_x such that $f(x) < n_x$. Therefore, $f_n(x) = |f(x)|$ for $n \geq n_x$. In particular, the sequence (f_n) converges to $|f|$ a.e. Moreover, for every x, $(f_n(x))$ is an increasing sequence (why?). Hence, the MCT applies and

$$\lim_n \int f_n d\mu = \int |f| d\mu.$$

Thus, for every $\epsilon > 0$ we can find n_0 such that

$$0 \leq \int |f| d\mu - \int f_{n_0} d\mu < \frac{\epsilon}{2}.$$

Let $0 < \delta < \frac{\epsilon}{2n_0}$ and let A be a measurable set such that $\mu(A) < \delta$. Then,

$$\int_A |f| d\mu = \int_A (|f| - f_{n_0}) d\mu + \int_A f_{n_0} d\mu.$$

Note now that

$$0 \leq \int_A (|f| - f_{n_0}) d\mu \leq \int (|f| - f_{n_0}) d\mu < \frac{\epsilon}{2},$$

and

$$\int_A f_{n_0} d\mu \leq n_0 \mu(A) < n_0 \delta < \frac{\epsilon}{2}.$$

Therefore,

$$\int_A |f| d\mu = \int_A (|f| - f_{n_0}) d\mu + \int_A f_{n_0} d\mu < \frac{\epsilon}{2} + \frac{\epsilon}{2} = \epsilon.$$

This completes the proof of Lemma 3.2.

We are now ready for another sufficient condition for uniform integrability.

Proposition 3.3 *Let (f_n) be a sequence of measurable functions. Let g be an integrable function. If for every $n \geq 1$, $|f_n| \leq g$ a.e., then the sequence (f_n) is uniformly integrable.*

Proof of Proposition 3.3
For every $n \geq 1$ and $k > 0$ we have

$$|f_n| \mathbf{1}_{\{|f_n| > k\}} \leq g \mathbf{1}_{\{g > k\}},$$

and integrating both sides of the inequality yields

$$\int_{\{|f_n| > k\}} |f_n| d\mu \leq \int_{\{g > k\}} g d\mu.$$

Hence,

$$\sup_n \int_{\{|f_n| > k\}} |f_n| d\mu \leq \int_{\{g > k\}} g d\mu.$$

By Lemma 3.1, the r.h.s. goes to 0 as k goes to $+\infty$. Therefore,

$$\lim_{k\to+\infty} \sup_n \int_{\{|f_n|>k\}} |f_n| d\mu = 0.$$

This completes the proof of Proposition 3.3.

The next theorem explains the importance of uniform integrability in Lebesgue theory.

Theorem 3.1 *Assume that $\mu(X) < +\infty$. Consider a sequence (f_n) in $L^1(\mu)$ and let f be in $L^1(\mu)$. Assume that the sequence (f_n) converges in measure to f. Then, (f_n) converges to f in $L^1(\mu)$ (i.e., $\lim_n \int |f_n - f| d\mu = 0$) if and only if the sequence (f_n) is uniformly integrable.*

Hence, under rather general hypotheses (finite measure space and convergence in measure) uniform integrability is a necessary and sufficient condition for convergence in $L^1(\mu)$! Note that the results we have seen so far such as the MCT or the DCT are only sufficient conditions for a mode of convergence that is weaker than convergence in $L^1(\mu)$.

Proof of Theorem 3.1
We first show that if (f_n) converges in $L^1(\mu)$, then (f_n) is uniformly integrable.
Let $\epsilon > 0$, there exists N such that if $n \geq N$, then

$$\int |f_n - f| d\mu < \frac{\epsilon}{2}. \tag{3.1}$$

By Lemma 3.2 for every $n \geq 1$ there exists a $\delta_n > 0$ such that if A is measurable and $\mu(A) < \delta_n$, then

$$\int_A |f_n| d\mu < \frac{\epsilon}{2}.$$

Similarly, there is a $\delta_0 > 0$ such that $\int_A |f| d\mu < \frac{\epsilon}{2}$. Let $\delta = \min(\delta_0, \delta, \dots, \delta_N)$, then $\delta > 0$ (why?). If $\mu(A) < \delta$, then

$$\int_A |f_n| d\mu < \frac{\epsilon}{2} \text{ for } 1 \leq n \leq N, \tag{3.2}$$

and

$$\int_A |f| d\mu < \frac{\epsilon}{2}. \tag{3.3}$$

Since the sequence (f_n) converges in the metric space $L^1(\mu)$ it must be bounded (see Proposition 2.4 in Chapter 7). Hence,

$$\sup_n \int |f_n| d\mu < +\infty.$$

Therefore, we can pick $k_0 > 0$ large enough to have

$$\frac{1}{k_0} \sup_n \int |f_n| d\mu < \delta. \tag{3.4}$$

Observe now that

$$\int |f_n| d\mu \geq \int_{\{|f_n| > k_0\}} |f_n| d\mu \geq k_0 \mu(|f_n| > k_0).$$

Using the last inequality and (3.4), we get

$$\mu(|f_n| > k_0) \leq \frac{1}{k_0} \int |f_n| d\mu < \delta \text{ for all } n \geq 1. \tag{3.5}$$

Since $\mu(|f_n| > k_0) < \delta$, we have by (3.2)

$$\int_{\{|f_n| > k_0\}} |f_n| d\mu < \frac{\epsilon}{2} \text{ for } 1 \leq n \leq N. \tag{3.6}$$

We now turn to the case $n > N$. By the triangle inequality,

$$\int_{\{|f_n| > k_0\}} |f_n| d\mu \leq \int_{\{|f_n| > k_0\}} |f_n - f| d\mu + \int_{\{|f_n| > k_0\}} |f| d\mu. \tag{3.7}$$

By (3.1) we have for $n > N$

$$\int_{\{|f_n| > k_0\}} |f_n - f| d\mu \leq \int |f_n - f| d\mu < \frac{\epsilon}{2}. \tag{3.8}$$

Using again that $\mu(|f_n| > k_0) < \delta$, we have by (3.3)

$$\int_{\{|f_n| > k_0\}} |f| d\mu < \frac{\epsilon}{2}. \tag{3.9}$$

Using (3.8) and (3.9) in (3.7) yields for $n > N$

$$\int_{\{|f_n| > k_0\}} |f_n| d\mu \leq \frac{\epsilon}{2} + \frac{\epsilon}{2} = \epsilon. \tag{3.10}$$

Putting together (3.6) and (3.10) we see that for any $\epsilon > 0$ we have found k_0 such that

$$\int_{\{|f_n|>k_0\}} |f_n| d\mu < \epsilon \text{ for all } n \geq 1.$$

That is, the sequence (f_n) is uniformly integrable.

We now prove the converse of Theorem 3.1.

We assume that (f_n) converges to f in measure and (f_n) is uniformly integrable. For $k > 0$, let the function ϕ_k be defined by

$$\begin{aligned}
\phi_k(x) =& k \quad \text{if } x \geq k \\
\phi_k(x) =& x \quad \text{if } |x| < k \\
\phi_k(x) =& -k \text{ if } x \leq -k
\end{aligned}$$

By the triangle inequality we have

$$|f_n - f| \leq |f_n - \phi_k(f_n)| + |\phi_k(f_n) - \phi_k(f)| + |\phi_k(f) - f|.$$

Integrating the inequality we get

$$\int |f_n - f| d\mu \leq \int |f_n - \phi_k(f_n)| d\mu + \int |\phi_k(f_n) - \phi_k(f)| d\mu + \int |\phi_k(f) - f| d\mu.$$

It is easy to check that for every real x we have

$$|\phi_k(x) - x| \leq |x| \mathbf{1}_{\{|x|>k\}}.$$

Therefore, for $n \geq 1$ and $k > 0$

$$\int |f_n - f| d\mu \leq \int_{\{|f_n|>k\}} |f_n| d\mu + \int |\phi_k(f_n) - \phi_k(f)| d\mu + \int_{\{|f|>k\}} |f| d\mu.$$

Let $\epsilon > 0$. By the uniform integrability of the sequence there exists k_1 such that if $k > k_1$

$$\int_{\{|f_n|>k\}} |f_n| d\mu < \frac{\epsilon}{3},$$

for all $n \geq 1$. Since f is integrable, by Lemma 3.1 there exists a $k_2 > 0$ such that if $k > k_2$

$$\int_{\{|f|>k\}} |f| d\mu < \frac{\epsilon}{3}.$$

We now pick a $k > \max(k_1, k_2)$ and turn to $\int |\phi_k(f_n) - \phi_k(f)| d\mu$.

Observe that for all real numbers x and y we have

$$|\phi_k(x) - \phi_k(y)| \leq |x - y|.$$

Using this inequality we see that since (f_n) converges in measure to f, then $(\phi_k(f_n))$ converges in measure to $\phi_k(f)$ (why?). Moreover, for every $n \geq 1$

$$|\phi_k(f_n)| \leq k.$$

Since $\mu(X) < +\infty$ the constant k is integrable. Hence, the Bounded Convergence Theorem (Theorem 2.1) applies to the sequence $(\phi_k(f_n))$. That is, there exists a natural N_1 such that for $n \geq N_1$

$$\int |\phi_k(f_n) - \phi_k(f)| d\mu < \frac{\epsilon}{3}.$$

Putting together our three estimates we get that for $n \geq N_1$

$$\begin{aligned}\int |f_n - f| d\mu &\leq \int_{\{|f_n|>k\}} |f_n| d\mu + \int |\phi_k(f_n) - \phi_k(f)| d\mu + \int_{\{|f|>k\}} |f| d\mu \\ &< \frac{\epsilon}{3} + \frac{\epsilon}{3} + \frac{\epsilon}{3} \\ &= \epsilon\end{aligned}$$

This proves that (f_n) converges to f in $L^1(\mu)$. The proof of Theorem 3.1 is complete.

Problems

1. Assume that (f_n) converges in $L^1(\mu)$. Show that

$$\sup_n \int |f_n| d\mu < +\infty.$$

2. Assume $\mu(X) < +\infty$. Give a new proof of the Dominated Convergence Theorem using Proposition 3.3 and Theorem 3.1.

3. Consider the Lebesgue measure m on the real line. For $n \geq 1$, let $f_n = \frac{1}{n}\mathbf{1}_{(0,n)}$.

(a) Show that (f_n) is uniformly integrable.
(b) Show that (f_n) converges in measure.
(c) Show that (f_n) does not converge in $L^1(m)$.

4. Let $\phi : \mathbb{R} \longrightarrow [0, +\infty)$ be such that

$$\lim_{x \to \infty} \frac{x}{\phi(x)} = 0.$$

Let (f_n) be a sequence of measurable functions such that

$$\sup_n \int \phi(|f_n|) d\mu < +\infty.$$

Show that the sequence (f_n) is uniformly integrable.

References

T.M. Apostol, *Calculus* (Blaisdell Publishing Company, New York, 1961)
B.R. Choe, An elementary proof of $\sum_{n=1}^{\infty} 1/n^2 = \pi^2/6$. Am. Math. Monthly **94**, 662–663 (1987)
W. Dunham, *The Calculus Gallery* (Princeton University Press, Princeton, 2005)
W. Feller, *An Introduction to Probability Theory and Its Applications*, vol. II, 2nd edn. (Wiley, New York, 1971)
P.M. Fitzpatrick, *Advanced Calculus*, 2nd edn. (American Mathematical Society, Providence, 2006)
G.B. Folland, *Real Analysis*, 2nd edn. (Wiley-Interscience, Hoboken, 1999)
S. Krantz, *Real Analysis and Foundations* (CRC Press, Boca Raton, 1991)
E. Kreiszig, *Introductory Functional Analysis with Applications* (Wiley, New York, 1978)
W. Rudin, *Principles of Mathematical Analysis*, 3rd edn. (McGraw Hill, New York, 1976)
W. Rudin, *Real and Complex Analysis*, 3rd edn. (McGraw Hill, New York, 1987)
R.B. Schinazi, *From Calculus to Analysis* (Birkhauser, Boston, 2011)

R. B. Schinazi, *From Classical to Modern Analysis*,
https://doi.org/10.1007/978-3-319-94583-5

Index

A
Abel's convergence test, 69
Abel's Theorem, 107
Almost everywhere, 212
Apostol, T.M., 1, 2
Archimedean property, 11

B
Baire's Category Theorem, 92
Bernstein polynomials, 88
Bolzano-Weierstrass Theorem, 29
Borel-Cantelli Lemma, 188
Borel measurable, 174
Borel σ-algebra, 169
Bounded Convergence Theorem, 253

C
Cantor set, 150, 191
Cauchy condensation test, 58
Cauchy convergence criterion for series, 64
Cauchy criterion for uniform convergence, 82
Cauchy-Schwarz inequality, 116
Cauchy sequence, 32, 125
Chebyshev's inequality, 211
Choe, B.R., 110
Closed ball, 142
Closed set, 140
Closure of a set, 142
Compact set, 146
Complete measure space, 225
Complete metric space, 127
Continuity, 155
Counting measure, 184

D
de Morgan's laws, 140
Density, 16, 135, 247
Dominated Convergence Theorem, 221
Dunham, W., 97

E
Egoroff's Theorem, 255
Equivalent metrics, 133
Euler's formula, 110

F
Fatou's Lemma, 210
Feller, W., 88
Folland, G.B., 172, 190

I
Image, 157
Indicator function, 175
Interior of a set, 145
Intervals, 163
Inverse image, 157

K
Krantz, S., 1, 92, 171
Kreiszig, E., 132, 150

L
Lebesgue integrable, 229
Lebesgue measure, 189
Lebesgue σ-algebra, 189
Limit inferior, 44

R. B. Schinazi, *From Classical to Modern Analysis*,
https://doi.org/10.1007/978-3-319-94583-5

Limit superior, 44
Littlewood's principles, 192
Lusin's Theorem, 192

M
Measurable function, 174
Measurable set, 167
Measure space, 183
Metric space, 115
Minkowski's inequality, 116
Monotone convergence theorem, 206
M-Weierstrass test, 93

N
Nowhere continuous function, 175
Nowhere differentiable function, 97
Null set, 190

O
Open ball, 137
Open set, 137

P
Partition, 197
Pointwise convergence, 77

R
Rearrangement of a series, 71
Riemann integral, 235
Riemann integral metric, 130
Rudin, W., 72, 247

S
Schinazi, R.B., 14, 71, 83, 89, 104, 109, 110, 119
σ-algebra, 167
Stirling's Formula, 109

U
Uniform continuity, 88, 161
Uniform convergence, 79
Uniform integrability, 257
Uniformly cauchy, 82